Birkhäuser

Operator Theory: Advances and Applications

Volume 212

Founded in 1979 by Israel Gohberg

Elementary Operators and Their Applications

3rd International Workshop
held at Queen's University Belfast,
14–17 April 2009

Raúl E. Curto
Martin Mathieu
Editors

Birkhäuser

Editors
Raúl E. Curto
Department of Mathematics
University of Iowa
14 MacLean Hall
Iowa City, IA 52242-1419
USA
raul-curto@uiowa.edu

Martin Mathieu
Department of Pure Mathematics
Queen's University Belfast
University Road
Belfast BT7 1NN
Northern Ireland
m.m@qub.ac.uk

2010 Mathematics Subject Classification: Primary 47B47; Secondary 46-06, 46A32, 46H35, 46H99, 46L05, 46L07, 47-06, 47A05, 47A10, 47A13, 47A62, 47B05

ISBN 978-3-0348-0036-5 e-ISBN 978-3-0348-0037-2
DOI 10.1007/978-3-0348-0037-2

Library of Congress Control Number: 2011920820

Cover design: deblik, Berlin

Printed on acid-free paper

Springer Basel AG is part of Springer Science+Business Media

www.birkhauser-science.com

Contents

Introduction

The *Third International Workshop on Elementary Operators and Their Applications* was held at the Department of Pure Mathematics of Queen's University Belfast between April 14 and 17, 2009. It was organised by Dr Martin Mathieu as a satellite to the joint meeting of the *British Mathematical Colloquium* and the *Annual Meeting of the Irish Mathematical Society* that took place at NUI Galway the week before. The funding received from the Irish Mathematical Society and the London Mathematical Society is gratefully acknowledged.

This series of workshops started in June 1991 at the Heinrich Fabri-Institute of the University of Tübingen, Germany and was continued in September 2001 at the University of Helsinki, Finland. At each of these meetings substantial progress in the research on elementary operators was reported. The proceedings volume of the 1991 workshop listed a number of difficult problems some of which, notably the norm problem for elementary operators have been solved in the meantime.

The present volume once again aims to present the state-of-the-art of our understanding of the theory of elementary operators and their applications. In fact, the applications are far more wide-ranging than one might expect and touch on many areas in Mathematics but also in Physics, such as Solid State Physics and Quantum Information Theory. Elementary operators are so simple in their definition that they occur everywhere; for instance, every linear mapping on a finite-dimensional semisimple Banach algebra is an elementary operator. It is no surprise therefore that one tries to approximate various classes of operators in infinite dimensions by elementary operators; some of the contributions in this book discuss these.

Yet the simple definition of an elementary operator entirely conceals the intricate and often challenging interplay between structural properties of elementary operators and the underlying algebras. This volume contains solicited articles by speakers at the workshop ranging from expository surveys to original research papers, each of which carefully refereed. They all bear witness to the very rich mathematics that is connected with the study of elementary operators, may it be multivariable spectral theory, the invariant subspace problem or tensor products of C^*-algebras.

As always in mathematics, results lead to new questions, and therefore the volume concludes with another list of open problems some of which were explained in the workshop problem session. May they inspire further exciting insights on elementary operators!

31 July 2010

Raúl E Curto, Iowa City
Martin Mathieu, Belfast

A picture of the workshop participants taken by the organiser

List of Participants

Jerónimo Alaminos, Universidad de Granada, Granada, Spain
Rob Archbold, University of Aberdeen, Aberdeen, Scotland
Ariel Blanco, Queen's University Belfast, Belfast, Northern Ireland
Nadia Boudi, Université Moulay Ismail, Meknès, Morocco
Raúl Curto, University of Iowa, Iowa City, USA
Lawrence Fialkow, SUNY at New Paltz, New Paltz, USA
Robin Harte, Trinity College Dublin, Dublin, Ireland
Derek Kitson, Trinity College Dublin, Dublin, Ireland
Rupert Levene, Trinity College Dublin, Dublin, Ireland
Bojan Magajna, University of Ljubljana, Ljubljana, Slovenia
Martin Mathieu, Queen's University Belfast, Belfast, Northern Ireland
Martin McGarvey, Queen's University Belfast, Belfast, Northern Ireland
F. Javier Meri, Granada, Universidad de Granada, Granada, Spain
Savvas Papapanagides, Queen's University Belfast, Belfast, Northern Ireland
Robert Pluta, Trinity College Dublin, Dublin, Ireland
Peter Rosenthal, University of Toronto, Toronto, Canada
Eero Saksman, University of Helsinki, Helsinki, Finland
Victor Shulman, Vologda State Technical University, Vologda, Russia
Franek Szafraniec, Jagiellonian University, Kraków, Poland
Richard Timoney, Trinity College Dublin, Dublin, Ireland
Ivan Todorov, Queen's University Belfast, Belfast, Northern Ireland
Ta Ngoc Tri, Lancaster University, Lancaster, England
Aleksej Turnšek, University of Ljubljana, Ljubljana, Slovenia
Lyudmila Turowska, Chalmers University of Technology and the University of Gothenburg, Gothenburg, Sweden
Hans-Olav Tylli, University of Helsinki, Helsinki, Finland
Armando R. Villena, Universidad de Granada, Granada, Spain

3rd International Workshop on Elementary Operators and their Applications (ElOp2009)

Programme

	Tuesday, 14 April	Wednesday, 15 April	Thursday, 16 April	Friday, 17 April
9:30 – 10:20		*Rob Archbold*	*Bojan Magajna*	*Lyudmila Turowska*
10:30 – 11:00		*Robin Harte*	*Martin McGarvey*	*Problem Session*
11:00 – 11:30		Coffee Break	Coffee Break	Coffee Break
11:30 – 12:20		*Richard Timoney*	*Nadia Boudi*	*Hans-Olav Tylli*
12:30 – 14:00	***Registration***	Lunch Break	Lunch Break	Lunch
14:00 – 14:50	Welcome	*Raúl Curto*	*Larry Fialkow*	***End of workshop***
	Victor Shulman			
15:00 – 15:30	*Ivan Todorov*	*Franek Szafraniec*	*Derek Kitson*	
15:30 – 16:00	Coffee Break	Coffee Break	Coffee Break	
16:00 – 16:50	*Ariel Blanco*		*Aleksej Turnsek*	
17:00 – 17:50	*Peter Rosenthal*	Football Match *Queen's PEC* **Or** Black Taxi Tour through Belfast	*Eero Saksman*	
18:00 – 19:00	***Reception***			
19:30			***Conference Dinner***	

Abstracts of Talks

States, representations and norms of elementary operators

Rob Archbold
University of Aberdeen
`r.archbold@maths.abdn.ac.uk`
Coauthors: Douglas Somerset (Aberdeen) and Richard Timoney (Dublin)

We discuss recent work with Douglas Somerset and Richard Timoney which aims to localize the connection between matricial norms of elementary operators on a C^*-algebra and the weak*-approximation of factorial states.

Let T be an elementary operator on a C^*-algebra A, π an irreducible representation of A and T^π the induced operator on $\pi(A)$. It is automatic that if n is a positive integer then $\|T^\pi\|_n \leq \|T\|_n$ and that the n-positivity of T implies the n-positivity of T^π. We show that if the upper multiplicity $M_U(\pi) > 1$ then $\|T^\pi\|_k \leq \|T\|_n$ for certain values of $k > n$ and that the n-positivity of T implies the k-positivity of T^π for these same larger values of k.

The results are obtained by using Timoney's descriptions of matricial norms and k-positivity in terms of the "tracial geometric mean" and factorial states. The condition on $M_U(\pi)$ allows one to approximate factorial states associated with π by type I factorial states of lower degree.

These localizations at irreducible representations π lead to new proofs of various characterizations of the class of antiliminal-by-abelian C^*-algebras in terms of factorial states and elementary operators.

On the cohomology of Banach operator algebras

Ariel Blanco
Queen's University Belfast
`a.blanco@qub.ac.uk`

We present new results on the bounded cohomology of Banach operator algebras. Some of them will be generalizations of well-known results for properly infinite C^*-algebras.

On spectrally bounded elementary operators

Nadia Boudi
Université Moulay Ismail
nadia_boudi@hotmail.com

I will first recall basic results on spectrally bounded linear maps, and then I will describe some spectrally bounded elementary operators. At the end of my talk, I will discuss some related conditions.

2-variable weighted shifts built from elementary tensors

Raúl E. Curto
University of Iowa
rcurto@math.uiowa.edu
Coauthors: Sang Hoon Lee and Jasang Yoon

We consider the class $\mathcal{TC}$ of 2-variable weighted shifts with tensor core. These are shifts whose restrictions to a large invariant subspace split as

$$(I \otimes W_\alpha, W_\beta \otimes I),$$

so these restrictions are canonically associated to left and right multiplication operators.

For the class $\mathcal{TC}$, we study the Reconstruction-of-the-Measure Problem (ROMP), which consists of finding necessary and sufficient conditions on a pair $\mathbf{T} \equiv (T_1, T_2) \in \mathcal{TC}$ that guarantee its subnormality. ROMP is intimately connected to the Lifting Problem for Commuting Subnormals (LPCS), which asks for necessary and sufficient conditions for a pair of subnormal operators on Hilbert space to admit commuting normal extensions. It is well known that the commutativity of the pair is necessary but not sufficient, and it has recently been shown that the joint hyponormality of the pair is necessary but not sufficient. Moreover, while abstract solutions of LPCS exist, concrete solutions are only known in very specific situations.

Our previous research has shown that many of the nontrivial aspects of LPCS are best detected within the class $\mathfrak{H}_1$ of commuting hyponormal pairs of subnormal operators, so we focus our attention on this class. $\mathcal{TC}$ is a large subclass of $\mathfrak{H}_1$, and we have been able to provide a complete solution of LPCS within $\mathcal{TC}$. We also show that for a pair $\mathbf{T} \in \mathcal{TC}$, the subnormality of (T_1^2, T_2) (typically easier to verify) automatically implies the subnormality of $\mathbf{T}$.

Abstract vs. concrete solutions to multivariable truncated moment problems

Lawrence Fialkow
SUNY New Paltz
fialkowl@newpaltz.edu

In previous work with R.E. Curto, we formulated two solutions to the truncated multivariable moment problem, one in terms of "flat extensions" of positive moment matrices, another in terms of positive extensions of linear functionals on polynomial spaces associated with the moment data. These solutions are difficult to implement concretely. We discuss some recent concrete results concerning multivariable quadratic moment problems and the bivariatic quartic moment problem (joint work with Jiawang Nie), and a concrete solution to the $y = x^3$ truncated moment problem.

Hermitian subspaces and Fuglede operators (Contributed talk)

Robin Harte
Trinity College Dublin
rharte@maths.tcd.ie

A real-linear subspace $H \subseteq A$ of a complex linear algebra with identity 1 will be called a "Hermitian subspace" if it satisfies

$$H \cap iH = \{0\} \ ; \ 1 \in H \ .$$

There is induced an involution $* : h + ik \mapsto h - ik$ on the complex linear subspace $H + iH \subseteq A$. When $A = B(X)$ is the bounded linear operators on a Banach space an involution on $H + iH$ gives rise to "Fuglede" operators $T \in H + iH$ for which

$$T^{-1}(0) \subseteq T^{*-1}(0) \ .$$

There is also induced another involution on an appropriate subspace of the "elementary operators" on $A = B(X)$. In this discussion we relate the Fuglede property for T and $S^\dagger$ to Fuglede properties for $L_T - R_S$ and $L_T R_S$.

A multivariable spectral mapping theorem (Contributed talk)

Derek Kitson
Trinity College Dublin
dk@maths.tcd.ie

The classical notions of ascent and descent for a linear operator on a vector space can be extended to arbitrary collections of operators. Using this fact we construct a Browder joint spectrum for commuting n-tuples of bounded operators on a complex Banach space. This Browder joint spectrum is compact valued and satisfies a multivariable spectral mapping theorem. Connections to the Browder spectrum of an elementary operator will be discussed.

Approximation by elementary operators

Bojan Magajna
University of Ljubljana
bojan.magajna@fmf.uni-lj.si

On which C^*-algebras A can all complete contractions, that preserve closed two-sided ideals, be approximated by completely contractive elementary operators? In the operator norm topology this is possible only if A is a finite direct sum of homogeneous C^*-algebras arising from C^*-bundles of finite type. On the other hand, pointwise approximation is always possible if A is nuclear, but the precise characterization is still open.

Normalisers, nest algebras and tensor products (Contributed talk)

Martin McGarvey
Queens University Belfast
mmcgarvey07@qub.ac.uk

If $\mathcal{A}$ is an operator algebra acting on a Hilbert space H, a *normaliser* of $\mathcal{A}$ is an operator T on H such that $T^*\mathcal{A}T \subseteq \mathcal{A}$ and $T\mathcal{A}T^* \subseteq \mathcal{A}$. The set of all normalisers of $\mathcal{A}$ is denoted by $N(\mathcal{A})$.

We will show that if $\mathcal{A}$ is the tensor product of finitely many continuous nest algebras, $\mathcal{B}$ is a CDCSL algebra and $N(\mathcal{A}) = N(\mathcal{B})$ then either $\mathcal{A} = \mathcal{B}$ or $\mathcal{A} = \mathcal{B}^*$.

Some not-quite-elementary operators

Peter Rosenthal
University of Toronto
rosent@math.toronto.edu
Coauthors: Don Hadwin and Eric Nordgren

This will be a discussion of joint work with Don Hadwin and Eric Nordgren concerning operator equations such as $AXB + CYD = Z$.

Alexandrov measures and connectedness in the space of composition operators

Eero Saksman
University of Helsinki
saksman@mappi.helsinki.fi
Coauthors: Eva Gallardo (Zaragoza), Maria Gonzales (Cadiz) and Pekka Nieminen (Helsinki)

We show that compact composition operators on the Hardy space H^2 do not form a component in the norm topology of operators. This answers a question of J.H. Shapiro and C. Sundberg from the early 1990's.

On elementary operators with compact coefficients

Victor Shulman
Vologda State Technical University
shulman_v@yahoo.com
Coauthors: Yuri Turovskii

W. Wojtynsky proved in 1970 that each solution of a linear operator equation of the form $\sum_k A_k X B_k + AX + XB = \lambda X$, where A_k, B_k, A, B are compact operators on a Banach space and $\lambda \neq 0$, is a nuclear operator. We suggest a new approach to this result and extend it considering eigenvectors and spectral subspaces of elementary operators with sufficiently many compact coefficients and of operators of more general form (in particular infinite sums are admitted).

On Murphy's positive definite kernels and Hilbert C^*-modules (Contributed talk)

Franciszek Hugon Szafraniec
Jagiellonian University Krakow
umszafra@cyf-kr.edu.pl

The paper the title refers to is that in Proceedings of the Edinburgh Mathematical Society, 40 (1997), 367–374. Taking it as an excuse we intend to realize a twofold purpose: 1. to atomize that important result showing by the way connections which are out of favour; 2. to rectify a tiny piece of history. The objective 1 is going to be achieved by adopting means adequate to goals; it is of great gravity and this is just Mathematics. The other, 2, comes from the author's internal need of showing how ethical values in Mathematics are getting depreciated. The latter have nothing to do with the previous issue; the coincidence is totally accidental.

Computation versus formulae for norms of elementary operators

Richard Timoney
Trinity College Dublin
richardt@maths.tcd.ie

We survey long standing and newer results on norms of elementary operators, including those of Stampfli (on derivations) and Haagerup (on completely bounded norms). We consider them from the point of view of their effectiveness for a practical problem and their value as a theoretical device. We discuss some aspects where there may be scope for further progress.

s-numbers of elementary operators (Contributed talk)

Ivan Todorov
Queen's University Belfast
i.todorov@qub.ac.uk
Coauthors: M. Anoussis and V. Felouzis

A theorem of Fong and Sourour states that an elementary operator acting on $B(H)$ is compact if and only if it has a representation where its symbols are compact operators. In this talk, a quantitative version of this result will be presented where the behaviour of the s-numbers of an elementary operator is linked to the behaviour of the singular numbers of its symbols.

Orthogonality and Fuglede–Putnam theorem

Aleksej Turnsek
University of Ljubljana
aleksej.turnsek@fmf.uni-lj.si

We present some results on orthogonality of the range and the kernel of elementary operators and connect them with the Fuglede–Putnam theorem.

Beurling–Pollard type theorems

Lyudmila Turowska
Chalmers University Gothenburg
turowska@chalmers.se
Coauthors: Victor Shulman

We establish a version of Beurling–Pollard theorem for operator synthesis and apply it to derive some results on linear operator equations. In addition, we establish a Beurling–Pollard theorem for weighted Fourier algebra and use it to obtain ascent estimate for operators that are functions of generalized scalar operators. This is a joint work with Victor Shulman.

Two-sided multiplication operators on $L(L^p)$

Hans-Olav Tylli
University of Helsinki
hojtylli@cc.helsinki.fi

I will discuss the following results from [LST] and [JS] about the qualitative properties of the basic multiplication operators $L_A R_B$, where $S \mapsto ASB$ for fixed operators $A, B \in \mathcal{L}(L^p)$, on the space $\mathcal{L}(L^p)$ of the bounded operators on L^p.

(i) $L_A R_B$ is strictly singular $\mathcal{L}(L^p) \to \mathcal{L}(L^p)$ for $1 \leq p \leq \infty$ if and only if $A \neq 0$ and $B \neq 0$ are strictly singular operators on L^p, [LST],

(ii) $L_A R_B$ is weakly compact $\mathcal{L}(L^p) \to \mathcal{L}(L^p)$ for $2 < p < \infty$ if and only if either A is compact, B is compact, or $JA \in \overline{G_{\ell^2}}$ and $B \in \overline{G_{\ell^p}}$ for some isometry $J : L^p \to L^\infty$, [JS]. Here $A \in \mathcal{L}(L^p)$ belongs to the factorization ideal G_{ℓ^r} if $A = UV$, where $U \in \mathcal{L}(\ell^r, L^p)$ and $V \in \mathcal{L}(L^p, \ell^r)$.

References

[JS] W.B. Johnson and G. Schechtman, Multiplication operators on $L(L^p)$ and ℓ^p-strictly singular operators, J. Eur. Math. Soc. **10** (2008), 1105–1119.

[LST] M. Lindström, E. Saksman and H.-O. Tylli, Strictly singular and cosingular multiplications, Canad. J. Math. **57** (2005), 1249–1278.

Elementary Operators that are Spectrally Bounded

Nadia Boudi and Martin Mathieu

Abstract. We study spectrally bounded elementary operators of length two on a complex unital Banach algebra $\mathcal{A}$. Some related conditions are investigated as well. In particular, we show that if S is an elementary operator of length two such that $S(x)$ is quasi-nilpotent for every $x \in A$, then $S(x)^3 \in \mathrm{rad}\mathcal{A}$ for every $x \in \mathcal{A}$.

Mathematics Subject Classification (2000). 47B47; 46H99, 47B48.

Keywords. Banach algebras, elementary operators, spectrally bounded, finite spectrum.

1. Introduction

Let $\mathcal{A}$ be a unital complex Banach algebra with Jacobson radical rad$\mathcal{A}$. For $x \in \mathcal{A}$, let $r(x)$ denote the spectral radius of x. A linear mapping $T\colon \mathcal{A} \to \mathcal{A}$ is said to be *spectrally bounded* if there is $M > 0$ such that $r(Tx) \leq Mr(x)$ for every $x \in \mathcal{A}$ [14]. A systematic study of such operators was begun in [16], and since then many results, in particular on the structure of spectrally bounded operators and spectral isometries (i.e., spectral radius preserving mappings) have been obtained, see [9, 10, 11, 17, 18] to cite but a few. The paper [15] discusses some of the main open problems in this area.

As usual, L_a and R_a stand for the left and the right multiplication by $a \in \mathcal{A}$, respectively. Using the subharmonicity of the spectral radius Pták proved in [20] that L_a is spectrally bounded if and only if a commutes with every element in $\mathcal{A}$ modulo rad$\mathcal{A}$. (For an alternative argument, see [12].) Curto and Mathieu showed that the generalised inner derivation $L_a - R_b$, a, $b \in \mathcal{A}$ is spectrally bounded if and

This paper developed out of a collaboration started during the *3rd International Conference on Elementary Operators and their Applications* held at Queen's University Belfast in April 2009. The first-named author thanks the second-named author for the generous hospitality during this meeting. The support by the London Mathematical Society is gratefully acknowledged.

R.E. Curto and M. Mathieu (eds.), *Elementary Operators and Their Applications*, Operator Theory: Advances and Applications 212, DOI 10.1007/978-3-0348-0037-2_1,

only if both L_a and R_b are spectrally bounded [12, Theorem B], thus extending Brešar's earlier result for inner derivations $L_a - R_a$ from [7]. Little seems to be known in this direction about general elementary operators $S = \sum_{j=1}^{n} L_{a_j} R_{b_j}$ despite the fact that, for instance, the question when all $S \in \mathcal{E}\ell(\mathcal{A})$, the algebra of elementary operators on $\mathcal{A}$, are *spectrally infinitesimal* (that is, Sx is quasi-nilpotent for all $x \in \mathcal{A}$) is related to other properties of $\mathcal{A}$; see [19, Section 6] for example.

In this paper we make a start on studying elementary operators that are spectrally bounded by focussing on the case of length two. More precisely, we investigate the following three properties a linear mapping $T\colon \mathcal{A} \to \mathcal{A}$ can have:

(a) T is spectrally bounded;
(b) T is almost spectrally bounded;
(c) Tx has finite spectrum for all $x \in \mathcal{A}$.

It turns out that both (a) and (c) each imply (b) but clearly (a) $\not\Rightarrow$ (c) (just take the identity mapping) and (c) $\not\Rightarrow$ (a): left multiplication L_a with a a non-central element in a finite-dimensional semisimple Banach algebra $\mathcal{A}$ serves as an example. Operators with the property that every element in their range has finite spectrum have recently attracted some attention; see [8, 6, 5], for example. They appear naturally in our context, see Example 3.3 below. The finite spectrum condition often implies that the range (of the operator itself or a related one) is contained in the socle soc$\mathcal{A}$ of $\mathcal{A}$ whereas spectrally boundedness is related to having the range lying in the radical rad$\mathcal{A}$. It therefore comes as no surprise that some of our results are along those lines too, see Corollary 3.6 and Theorem 3.9.

An elementary operator leaves each ideal of $\mathcal{A}$ invariant; hence it is close at hand to use primitive ideals and dense algebras of operators to study properties of it. The Jacobson Density Theorem, and its variants, prevail as a major tool in our discussion too.

2. Length two elementary operators

Let $a, b \in \mathcal{A}$ and denote by $M_{a,b}$ the two-sided multiplication $x \mapsto axb$, $x \in \mathcal{A}$. Since $r(axb) = r(bax)$, it follows immediately from the aforementioned characterisation by Pták that $M_{a,b}$ is spectrally bounded if and only if $ba \in \mathcal{Z}(\mathcal{A})$, the centre modulo the radical of $\mathcal{A}$, which is defined by

$$\mathcal{Z}(\mathcal{A}) = \{z \in \mathcal{A} \mid zx - xz \in \mathrm{rad}\mathcal{A} \text{ for all } x \in \mathcal{A}\}.$$

A *length two* elementary operator is of the form $S = M_{a,b} + M_{c,d}$, where the sets $\{a, c\}$ and $\{b, d\}$ are linearly independent. (In fact, a stronger form of independence may have to be imposed in order to make it impossible to write such S as a length one operator $M_{u,v}$ depending on the properties of the centre of $\mathcal{A}$. For C^*-algebras, this is studied in detail in [2, Chapter 5], however for general Banach algebras the situation is far more complicated. Therefore, strictly speaking we consider elementary operators of length *at most two* but this slight ambiguity does not affect our results below.)

Without further assumptions, the two-sided multiplication $M_{a,b}$ is not a homomorphism; however, if ba is a non-zero scalar, $\phi = (ba)^{-1} M_{a,b}$ is easily checked to be multiplicative. This observation is the key to a sufficient criterion for spectrally boundedness of length two elementary operators in this section.

Lemma 2.1. *Let $\mathcal{A}$ be a unital Banach algebra, and let ϕ, ψ be two bounded homomorphisms on $\mathcal{A}$ such that $\psi(x)\phi(y) \in \operatorname{rad}\mathcal{A}$ for every $x, y \in \mathcal{A}$. Then, for all λ, $\mu \in \mathbb{C}$, the mapping $T = \lambda\phi + \mu\psi$ is spectrally bounded.*

Proof. Fix $x \in \mathcal{A}$. Take a non-zero complex number $\alpha \notin \sigma(\lambda x) \cup \sigma(\mu x)$. Then there exists $y \in \mathcal{A}$ such that

$$\frac{\lambda}{\alpha}\, x + y - y\frac{\lambda}{\alpha}\, x = 0$$

and hence

$$\frac{\lambda}{\alpha}\, \phi(x) + \phi(y) - \phi(y)\frac{\lambda}{\alpha}\, \phi(x) = 0.$$

Analogously, there exists $z \in \mathcal{A}$ such that

$$\frac{\mu}{\alpha}\, \psi(x) + \psi(z) - \psi(z)\frac{\mu}{\alpha}\, \psi(x) = 0.$$

We compute that

$$(1 - \phi(y) - \psi(z))\left(1 - \frac{\lambda}{\alpha}\, \phi(x) - \frac{\mu}{\alpha}\, \psi(x)\right) = 1 + s, \tag{2.1}$$

where $s = \frac{\mu}{\alpha}\, \phi(y)\psi(x) + \frac{\lambda}{\alpha}\, \psi(z)\phi(x)$. It follows from our assumption that $s^2 \in \operatorname{rad}\mathcal{A}$ and hence s is quasi-nilpotent. This implies that $1 + s$ is invertible which, together with identity (2.1), entails that $\alpha - (\lambda\phi + \mu\psi)(x)$ is left invertible. Since the boundary of the spectrum of Tx is contained in the left approximate point spectrum, it follows that $\alpha \notin \partial\sigma(Tx)$ and thus

$$\partial\sigma(Tx) \subseteq \sigma(\lambda x) \cup \sigma(\mu x).$$

This implies that $r(Tx) \leq \max\{|\lambda|, |\mu|\}\, r(x)$ for each $x \in \mathcal{A}$ which completes the proof. □

Lemma 2.2. *Let $\mathcal{A}$ be a unital Banach algebra. Let ψ be a bounded homomorphism on $\mathcal{A}$ and let ϕ be a linear mapping on $\mathcal{A}$ such that $\phi(x)^2 \in \operatorname{rad}\mathcal{A}$ for every $x \in \mathcal{A}$. Suppose that $\psi(x)\phi(y) \in \operatorname{rad}\mathcal{A}$ for all x, $y \in \mathcal{A}$ or $\phi(x)\psi(y) \in \operatorname{rad}\mathcal{A}$ for all x, $y \in \mathcal{A}$. Then the mapping $T = \phi + \psi$ is spectrally bounded.*

Proof. Let $x \in A$ and let α be a non-zero complex number such that $\alpha \notin \sigma(x)$. There exists $y_0 \in \mathcal{A}$ such that

$$\frac{\psi(x)}{\alpha} + \psi(y_0) - \psi(y_0)\frac{\psi(x)}{\alpha} = 0. \tag{2.2}$$

Suppose first that $\psi(x)\phi(y) \in \operatorname{rad}\mathcal{A}$ for all $x, y \in \mathcal{A}$. From identity (2.2) we obtain

$$(1 - \psi(y_0))\left(1 - \frac{\phi(x)}{\alpha} - \frac{\psi(x)}{\alpha}\right) = 1 - \frac{\phi(x)}{\alpha} + \psi(y_0)\frac{\phi(x)}{\alpha}. \tag{2.3}$$

As in the proof of Lemma 2.1 we deduce that $\alpha - Tx$ is left invertible and hence $r(Tx) \le r(x)$ for each x.

Suppose now that $\phi(x)\psi(y) \in \mathrm{rad}\mathcal{A}$ for all $x, y \in \mathcal{A}$. Identity (2.3) together with the fact that $\left(\frac{\phi(x)}{\alpha} - \psi(y_0)\frac{\phi(x)}{\alpha}\right)^2 \in \mathrm{rad}\mathcal{A}$ allow us to complete the proof as before. □

Throughout the following, $\mathrm{Prim}\mathcal{A}$ shall denote the set of primitive ideals of $\mathcal{A}$. An elementary operator $S \in \mathcal{E}\ell(\mathcal{A})$ leaves each ideal of $\mathcal{A}$ invariant. Therefore, for each $P \in \mathrm{Prim}\mathcal{A}$, it induces an elementary operator $S_P \in \mathcal{E}\ell(\mathcal{A}_P)$ on the primitive Banach algebra $\mathcal{A}_P = \mathcal{A}/P$. The well-known formula, [3, Theorem 4.2.1],

$$\sigma(x) = \bigcup_{P \in \mathrm{Prim}\mathcal{A}} \sigma(x_P) \qquad (x \in \mathcal{A}), \tag{2.4}$$

where $x_P = x + P$, enables us to reduce the situation to the primitive case: If each S_P is spectrally bounded and the spectral operator norms $\|S_P\|_\sigma$ (in the sense of [16]) are uniformly bounded, then S is spectrally bounded as

$$r(Sx) = \sup_P r(S_P x_P) \le \sup_P \|S_P\|_\sigma \, r(x_P) \le \sup_P \|S_P\|_\sigma \, r(x) \qquad (x \in \mathcal{A}).$$

Proposition 2.3. *Let $\mathcal{A}$ be a unital Banach algebra, and let $S = M_{a,b} + M_{c,d} \in \mathcal{E}\ell(\mathcal{A})$. Suppose that ba, $dc \in \mathcal{Z}(\mathcal{A})$ and $bc \in \mathrm{rad}\mathcal{A}$. Then S is spectrally bounded.*

Proof. Let P be a primitive ideal of $\mathcal{A}$. By hypothesis, $b_P c_P = 0$ and $b_P a_P = \lambda_P$, $d_P c_P = \mu_P$ are both scalars. If $\lambda_P = \mu_P = 0$ then we compute that $(S_P x_P)^3 = 0$ for each $x \in \mathcal{A}$ and thus $\|S_P\|_\sigma = 0$. If $\lambda_P \neq 0$ and $\mu_P \neq 0$ we put $\phi_P = \lambda_P^{-1} M_{a_P, b_P}$ and $\psi_P = \mu_P^{-1} M_{c_P, d_P}$ to obtain two (bounded and spectrally bounded) homomorphisms. Since $\psi_P(x_P)\phi_P(y_P) = 0$ for all $x, y \in \mathcal{A}$, we can apply Lemma 2.1 to conclude that $S_P = \lambda_P \phi_P + \mu_P \psi_P$ is spectrally bounded with $\|S_P\|_\sigma \le \max\{r(ba), r(dc)\}$. If, say, $\lambda_P = 0$ and $\mu_P \neq 0$ then the hypotheses of Lemma 2.2 are fulfilled, for $\phi_P = M_{a_P, b_P}$ and $\psi_P = \mu_P^{-1} M_{c_P, d_P}$, and we obtain $\|S_P\|_\sigma \le 1$. The discussion preceding this lemma show that S is spectrally bounded with $\|S\|_\sigma \le \max\{r(ba), r(dc), 1\}$. □

3. Main results

Clearly the condition on the coefficients in Proposition 2.3 is not necessary for S to be spectrally bounded; simply take $b = -c = 1$ and refer back to the results cited in the Introduction. We shall obtain a necessary and sufficient condition in Theorem 3.5 below. We shall also explore a weaker condition, n-almost spectrally boundedness in this section and obtain a necessary condition as well.

Our techniques are inspired by [12], in particular the use of Jacobson's Density Theorem and Sinclair's improvement on it, see [3, Theorem 4.2.5] and [3, Corollary 4.2.6], respectively. We shall also require the following variant of Amitsur's Lemma [1], which is obtained in [5, Lemma 2.1].

Lemma 3.1. *Let n, r be positive integers, U and V vector spaces over $\mathbb{C}$ and let V_0 be a finite-dimensional subspace of V. Let $T_1, \dots, T_n \colon U \to V$ be linear mappings. Suppose that, for every r-tuple of vectors $\zeta_1, \dots, \zeta_r$ in U, the set $\{T_1\zeta_i, \dots, T_n\zeta_i : i = 1, \dots, r\}$ is linearly dependent modulo V_0. Then there exists a non-trivial linear combination of $T_1, \dots, T_n$ of finite rank.*

Let X be a non-empty set. As usual, the cardinality of X will be denoted by $\#X$. Throughout, $\mathcal{A}$ is a unital Banach algebra and $n \geq 1$ a natural number.

Definition 3.2. Let $T \colon \mathcal{A} \to \mathcal{A}$ be a linear mapping. We say that T is *n-almost spectrally bounded* if there exists $M > 0$ such that, for every $x \in \mathcal{A}$,

$$\#\{\lambda \in \sigma(Tx) : |\lambda| > Mr(x)\} \leq n. \tag{3.1}$$

In this case, M is called *an n-almost spectral bound for T*.

Clearly, every spectrally bounded linear mapping is n-almost spectrally bounded for each n. Here is a less trivial example.

Example 3.3. Let $T \colon \mathcal{A} \to \mathcal{A}$ be a bounded linear mapping satisfying $\#\sigma(Tx) < \infty$ for each $x \in \mathcal{A}$. It follows from [8, Lemma 2.1] that there exists $n \in \mathbb{N}$ such that $\#\sigma(Tx) \leq n$ for all $x \in \mathcal{A}$. Thus T is n-almost spectrally bounded.

For a Banach space X, let $\mathscr{B}(X)$ denote the algebra of all bounded linear operators on X, and $\mathscr{F}(X)$ denote the ideal of all finite rank operators in $\mathscr{B}(X)$. Set $\mathscr{F}'(X) = \mathscr{F}(X) + \mathbb{C}I$, where I is the identity mapping.

Proposition 3.4. *Suppose that $S = M_{a,b} + M_{c,d} \in \mathscr{E}\ell(\mathcal{A})$ is n-almost spectrally bounded for some $n \in \mathbb{N}$. Then, for every irreducible representation π of $\mathcal{A}$ on a Banach space X, there exists $\beta \in \mathbb{C}$ such that*

$$\pi((b + \beta d)a) \in \mathscr{F}'(X) \quad \textit{and} \quad \pi(d(c - \beta a)) \in \mathscr{F}'(X)$$

and either $\pi(b + \beta d)\pi(c - \beta a)$ has finite rank or $\beta = 0$ and $\pi(da)$ has finite rank.

Proof. Let π be an irreducible representation of $\mathcal{A}$ on a Banach space X. In the case when X is finite dimensional, there is nothing to prove; so we shall assume that X is infinite dimensional. Suppose first that there exist $r \in \mathbb{N}$ and $\zeta_1, \dots, \zeta_r \in X$ such that the set $\{\zeta_j, \pi(ba)\zeta_j, \pi(da)\zeta_j : 1 \leq j \leq r\}$ is linearly independent. By [3, Corollary 4.2.6], for every $k \in \mathbb{N}$, we can find $x_k \in \exp \mathcal{A}$ such that, for each j, span $\{\zeta_j, \pi(ba)\zeta_j, \pi(da)\zeta_j\}$ is invariant under $\pi(x_k)$ and the corresponding matrix representation of $\pi(x_k)$ with respect to $\{\zeta_j, \pi(ba)\zeta_j, \pi(da)\zeta_j\}$ is

$$\begin{pmatrix} 0 & j & 0 \\ \frac{1}{k} & 0 & 0 \\ 0 & 0 & 1 \end{pmatrix}.$$

Choose $x \in \mathcal{A}$ with the property that $\pi(x)\zeta_j = \pi(ba)\zeta_j$ and $\pi(xda)\zeta_j = 0$, for $1 \leq j \leq r$. Set $x'_k = x_k x x_k^{-1}$. Then

$$\pi\big(S(x'_k)a\big)\zeta_j = jk\,\pi(a)\zeta_j$$

so that $\{k,\dots,rk\} \subseteq \sigma(\pi S(x'_k)) \subseteq \sigma(S(x'_k))$. Let M be an n-almost spectral bound for S. If $r > n$, then, for each $k \in \mathbb{N}$, we must have $k \le Mr(x)$ which is impossible. It follows that $r \le n$ and thus the set

$$\{\zeta_j,\, \pi(ba)\zeta_j,\, \pi(da)\zeta_j : 1 \le j \le n+1\}$$

is linearly dependent for all vectors $\zeta_1,\dots,\zeta_{n+1} \in X$. By Lemma 3.1, there exists a non-trivial linear combination of I, $\pi(ba)$ and $\pi(da)$ which has finite rank.

By symmetry, some non-trivial linear combination of I, $\pi(dc)$ and $\pi(bc)$ has finite rank.

Suppose first that $\pi(ba) \notin \mathscr{F}'(X)$ and $\pi(dc) \notin \mathscr{F}'(X)$ which allows us to write

$$\pi(da+\alpha ba) = \lambda I + F \quad \text{and} \quad \pi(bc+\beta dc) = \lambda' I + F',$$

where $\lambda,\lambda',\alpha,\beta \in \mathbb{C}$ and $F,F' \in \mathscr{F}(X)$.

Assume towards a contradiction that, for each $r \in \mathbb{N}$, there exist $\zeta_1,\dots,\zeta_r \in X$ such that the set

$$\{\zeta_j,\, \pi(ba)\zeta_j,\, \pi(dc)\zeta_j : 1 \le j \le r\}$$

is linearly independent modulo $FX + F'X$. For $k \in \mathbb{N}$, we can find $x_k \in \exp \mathcal{A}$ such that, for each j, span $\{\zeta_j, \pi(ba)\zeta_j, \pi(dc)\zeta_j\}$ is invariant under $\pi(x_k)$, the corresponding matrix representation of $\pi(x_k)$ with respect to $\{\zeta_j, \pi(ba)\zeta_j, \pi(dc)\zeta_j\}$ is

$$\begin{pmatrix} 0 & j & 0 \\ \frac{1}{k} & 0 & 0 \\ 0 & 0 & 1 \end{pmatrix}$$

and $\pi(x_k)_{|FX+F'X} = I$. Choose $x \in \mathcal{A}$ with the property that $\pi(x)\zeta_j = \pi(ba)\zeta_j$, $\pi(xba)\zeta_j = \pi(xdc)\zeta_j = 0$ and $\pi(x)F = \pi(x)F' = 0$, for $1 \le j \le r$. Set $x'_k = x_k x x_k^{-1}$. Then

$$\pi\big(S(x'_k)a\big)\zeta_j = jk\,\pi(a-\alpha c)\zeta_j \quad \text{and} \quad \pi\big(S(x'_k)c\big)\zeta_j = 0$$

so that $\{k,\dots,rk\} \subseteq \sigma(\pi S(x'_k))$. As above, we deduce that the set

$$\{\zeta_j,\, \pi(ba)\zeta_j,\, \pi(dc)\zeta_j : 1 \le j \le n+1\}$$

is linearly dependent modulo $FX+F'X$ for all vectors $\zeta_1,\dots,\zeta_{n+1} \in X$. Applying once again Lemma 3.1, we infer that there exists a non-trivial linear combination of I, $\pi(ba)$ and $\pi(dc)$, which has finite rank. Write

$$\pi(dc+\delta ba) = \gamma I + G,$$

for some $\delta,\gamma \in \mathbb{C}$ and $G \in \mathscr{F}(X)$.

Next choose vectors $\zeta_1,\dots,\zeta_r \in X$ such that the set $\{\zeta_j,\, \pi(ba)\zeta_j : 1 \le j \le r\}$ is linearly independent modulo $FX + F'X + GX$ which is possible since $\pi(ba) \notin \mathscr{F}'(X)$. For every $k \in \mathbb{N}$, we can find $x_k \in \exp \mathcal{A}$ such that, for each j, span $\{\zeta_j, \pi(ba)\zeta_j\}$ is invariant under $\pi(x_k)$, the corresponding matrix representation of $\pi(x_k)$ with respect to $\{\zeta_j,\, \pi(ba)\zeta_j\}$ is

$$\begin{pmatrix} 0 & 1 \\ \frac{1}{jk} & 0 \end{pmatrix}$$

and $\pi(x_k)_{|FX+F'X+GX} = I$. Choose $x \in \mathcal{A}$ with the property that $\pi(x)\zeta_j = \pi(ba)\zeta_j$, $\pi(xba)\zeta_j = \pi(ba)\zeta_j$, and $\pi(x)F = \pi(x)F' = \pi(x)G = 0$. Set $x'_k = x_k x x_k^{-1}$. Then the corresponding matrix representation of $\pi S(x'_k)$ with respect to $\{\pi(a)\zeta_j, \pi(c)\zeta_j\}$ is

$$\begin{pmatrix} jk & \lambda' - \beta\gamma + \beta\delta jk \\ \lambda - \alpha jk & \gamma - \delta jk \end{pmatrix}.$$

The trace of this matrix is $(1-\delta)jk + \gamma$. If $\delta \neq 1$ we can choose k sufficiently big to obtain a contradiction as above. Thus $\delta = 1$. The characteristic polynomial of the above matrix is

$$\chi(t) = t^2 - \gamma t - \lambda(\lambda' - \beta\gamma) + (jk)^2(\alpha\beta - 1) + jk(\alpha\lambda' + \gamma - \lambda\beta - \alpha\beta\gamma)$$

and has to be independent of k; thus we must have $\alpha\beta = 1$ and $\lambda' = \lambda\beta^2$. Consequently, $\pi(dc) + \pi(ba) \in \mathscr{F}'(X)$ and

$$\beta\pi(da) + \pi(ba) = \beta\lambda I + \beta F \quad \text{and} \quad \alpha\pi(bc) + \pi(dc) = \beta\lambda I + \alpha F'$$

from which it follows that $\pi(b + \beta d)\pi(c - \beta a) \in \mathscr{F}(X)$.

Suppose now that $\pi(ba) \in \mathscr{F}'(X)$ or $\pi(dc) \in \mathscr{F}'(X)$. Observe that, for every $\alpha \in \mathbb{C}$, we have

$$S = M_{a+\alpha c,b} + M_{c,d-\alpha b} = M_{a,b-\alpha d} + M_{c+\alpha a,d}. \tag{3.2}$$

By choosing a suitable α, we will reduce this situation to the previous one. We distinguish two cases:

Case 1: $\pi(ba) \in \mathscr{F}'(X)$ and $\pi(da) \notin \mathscr{F}'(X)$, or $\pi(dc) \in \mathscr{F}'(X)$ and $\pi(bc) \notin \mathscr{F}'(X)$. Suppose, for instance, the first possibility. Then, for at most one complex numbers α, we have $\pi(d(c + \alpha a)) \notin \mathscr{F}'(X)$ and $\pi((b - \alpha d)a) \notin \mathscr{F}'(X)$. Using identity (3.2) and the above steps, we get the desired result. More precisely, we show that $\pi(dc) \in \mathscr{F}'(X)$ and $\pi(bc) \in \mathscr{F}(X)$.

Case 2: $\pi(ba), \pi(da) \in \mathscr{F}'(X)$ or $\pi(dc), \pi(bc) \in \mathscr{F}'(X)$. Assume, for instance, that $\pi(ba), \pi(da) \in \mathscr{F}'(X)$. Suppose first that $\pi(dc) \in \mathscr{F}'(X)$. We claim that either $\pi(da) \in \mathscr{F}(X)$ or $\pi(bc) \in \mathscr{F}'(X)$. If $\pi(da) \notin \mathscr{F}(X)$ we write

$$\pi(da) = \lambda I + F,\ \pi(dc) = \gamma I + F',\ \pi(ba) = \mu I + G,$$

where $\lambda, \gamma, \mu \in \mathbb{C}$ and $F, F', G \in \mathscr{F}(X)$. Without loss of generality, we may suppose that $\lambda = 1$. Suppose that there are vectors $\zeta_1, \dots, \zeta_r \in X$ such that $\{\zeta_j, \pi(bc)\zeta_j : 1 \le j \le r\}$ is linearly independent modulo $FX + F'X + GX$. For every $k \in \mathbb{N}$, we can find $x_k \in \exp \mathcal{A}$ such that, for each j, span$\{\zeta_j, \pi(bc)\zeta_j\}$ is invariant under $\pi(x_k)$, the corresponding matrix representation of $\pi(x_k)$ with respect to $\{\zeta_j, \pi(bc)\zeta_j\}$ is

$$\begin{pmatrix} 0 & 1 \\ \frac{1}{jk} & 0 \end{pmatrix}$$

and $\pi(x_k)_{|FX+F'X+GX} = I$. Choose $x \in \mathcal{A}$ with the property that $\pi(x)\zeta_j = \pi(bc)\zeta_j$, $\pi(xbc)\zeta_j = \pi(bc)\zeta_j$ and $\pi(x)F = \pi(x)F' = \pi(x)G = 0$ for all j. Set $x'_k =$

$x_k x x_k^{-1}$. Then the corresponding matrix representation of $\pi S(x'_k)$ with respect to $\{\pi(a)\zeta_j, \pi(c)\zeta_j\}$ is

$$\begin{pmatrix} \mu & jk \\ 1 & \gamma \end{pmatrix}.$$

As above, we show that there exists $r \in \mathbb{N}$ such that the set

$$\{\zeta_1, \ldots, \zeta_r, \pi(bc)\zeta_1, \ldots, \pi(bc)\zeta_r\}$$

is linearly dependent for all vectors $\zeta_1, \ldots, \zeta_r$ in X. Thus, $\pi(bc) \in \mathscr{F}'(X)$ and the claim is proved.

Since $\pi(bc)$, $\pi(dc)$, $\pi(ba)$ and $\pi(da)$ now all belong to $\mathscr{F}'(X)$, there exists a complex number β such that $\pi(b+\beta d)\pi(c-\beta a) \in \mathscr{F}(X)$, and the proof of this case is complete.

Suppose finally that $\pi(dc) \notin \mathscr{F}'(X)$. We already showed above that there exists $\beta \in \mathbb{C}$ such that $\pi(bc+\beta dc) \in \mathscr{F}'(X)$. Re-writing S as in (3.2) with $\alpha = -\beta^{-1}$ provided $\beta \neq 0$ and $\alpha = 1$ if $\beta = 0$, we see that we are in Case 1. Thus, this case cannot occur. $\square$

Given $P \in \operatorname{Prim}\mathcal{A}$, we shall denote by π_P an irreducible representation of $\mathcal{A}$ on a Banach space X_P with kernel P.

Theorem 3.5. *Let $\mathcal{A}$ be a unital Banach algebra. Then $S = M_{a,b} + M_{c,d} \in \mathscr{E}\ell(\mathcal{A})$ is spectrally bounded if and only if, for every primitive ideal P of $\mathcal{A}$, there exists $\beta_P \in \mathbb{C}$ such that*

$$\pi_P((b+\beta_P d)a) \in \mathbb{C}I \quad \text{and} \quad \pi_P(d(c-\beta_P a)) \in \mathbb{C}I \tag{3.3}$$

and either $\pi_P(b+\beta_P d)\pi_P(c-\beta_P a) = 0$ or $\beta_P = 0$ and $\pi_P(da) = 0$. In particular, $ba + dc \in \mathcal{Z}(\mathcal{A})$ in this case.

Proof. The "only if"-part is a particular case of the proof of Proposition 3.4. The space X can be finite dimensional, and at each step, we work with one $\zeta \in X$ instead of an r-tuple.

To show the sufficiency of the condition, write $\pi_P((b+\beta_P d)a) = \lambda_P I$ for some $\lambda_P \in \mathbb{C}$ and note that (3.3) implies that $u_P \in \mathbb{C}I$ for every P, where $u = ba + dc$. We shall show that the set $\{\lambda_P : P \in \operatorname{Prim}\mathcal{A}\} \subseteq \mathbb{C}$ is bounded. If $|\beta_P| < \frac{1}{2}$ then

$$|\lambda_P| \leq \|\pi_P(ba)\| + \tfrac{1}{2}\|\pi_P(da)\| \leq \|ba\| + \|da\|.$$

Otherwise $|\beta_P| \geq \frac{1}{2}$, and the identity $\pi_P(b+\beta_P d)\pi_P(c-\beta_P a) = 0$ entails that

$$\beta_P \pi_P((b+\beta_P d)a) = \pi_P((b+\beta_P d)c)$$

and thus

$$|\lambda_P| \leq 2\|\pi_P(bc)\| + \|\pi_P(dc)\| \leq 2\|bc\| + \|dc\|.$$

Consequently there is $M_1 > 0$ such that $|\lambda_P| \leq M_1$ for all $P \in \operatorname{Prim}\mathcal{A}$.

We can use this to infer that

$$\begin{aligned} r(\pi_P(d(c-\beta_P a))) &= r(\pi_P(ba+dc) - \pi_P((b+\beta_P d)a)) \\ &= r(\pi_P(u) - \lambda_P I) \\ &\leq r(u) + M_1 \end{aligned}$$

for all $P \in \mathrm{Prim}\mathcal{A}$. Therefore there is $M_2 > 0$ such that $r(\pi_P(d(c-\beta_P a))) \leq M_2$ for all P. Writing S_P once more as

$$S_P = M_{a_P, b_P+\beta_P d_P} + M_{c_P-\beta_P a_P, d_P} = M_{c_P,d_P} + M_{a_P,b_P}$$

we apply Proposition 2.3 to conclude that each S_P, $P \in \mathrm{Prim}\mathcal{A}$ is spectrally bounded. Moreover, the proof of that proposition shows that

$$\|S_P\|_\sigma \leq \max\{M_1, M_2, 1\} \qquad (P \in \mathrm{Prim}\mathcal{A})$$

from which we conclude that S itself is spectrally bounded. □

We draw the following consequence for spectrally infinitesimal elementary operators of length two.

Corollary 3.6. *Let $S = M_{a,b} + M_{c,d} \in \mathcal{El}(\mathcal{A})$. If $S(x)$ is quasi-nilpotent for every $x \in \mathcal{A}$, then $S(x)^3 \in \mathrm{rad}\mathcal{A}$ for every $x \in \mathcal{A}$.*

Proof. Let P be a primitive ideal of $\mathcal{A}$ and let $\pi = \pi_P$. According to Theorem 3.5 there exists $\beta \in \mathbb{C}$ such that

$$\pi(ba + \beta da) \in \mathbb{C}I \quad \text{and} \quad \pi(ba + dc) \in \mathbb{C}I,$$

and either $\pi(b+\beta d)\pi(c-\beta a) = 0$ or $\beta = 0$ and $\pi(da) = 0$. Writing S as $S = M_{a,b+\beta d} + M_{c-\beta a,d}$ and replacing b by $b+\beta d$ and c by $c-\beta a$, if necessary, we may assume that $\beta = 0$ and either $\pi(bc) = 0$ or $\pi(da) = 0$. Set $\pi(ba) = \lambda I$ and $\pi(dc) = \lambda' I$. We shall prove that $\lambda = \lambda' = 0$.

Suppose that $\pi(da) = 0$ and $\pi(a) \neq 0$ (otherwise we clearly have $\lambda = 0$). Take $x \in \mathcal{A}$ and $\zeta \in X$ such that

$$\pi(x)\zeta = \zeta \quad \text{and} \quad \pi(a)\zeta \neq 0.$$

Then $\pi(S(x)a)\zeta = \lambda\pi(a)\zeta$ wherefore $\lambda \in \sigma(S(x)) = \{0\}$. It follows that $\pi(S(x)a) = 0$ for all $x \in \mathcal{A}$. If $\lambda' \neq 0$ we can find $\zeta \in X \setminus \{0\}$, $x \in \mathcal{A}$ and $\mu \in \mathbb{C}$ such that $\pi(xbc)\zeta = \mu\zeta$ and $\pi(xdc)\zeta = \zeta$. It follows that $\pi(S(x)c)\zeta = \mu\pi(a)\zeta + \pi(c)\zeta$. Upon applying $\pi(S(x))$ to this identity we arrive at a contradiction since $\pi(S(x)a) = 0$. The other possibility, i.e., $\pi(bc) = 0$, is treated analogously.

In either case we conclude that, for every $x \in \mathcal{A}$, $\pi S(x)^3 = 0$. As a result, $S(x)^3 \in \mathrm{rad}\mathcal{A}$ for every $x \in \mathcal{A}$. □

It follows from the above proof that $ba + dc \in \mathrm{rad}\mathcal{A}$ if $S(x)$ is quasi-nilpotent for every $x \in \mathcal{A}$.

In the remainder of this article we focus our attention on elementary operators of length two such that every element in their range has finite spectrum. As observed above (Example 3.3) such an operator is n-almost spectrally bounded, where n is a uniform bound for the number of scalars in the spectrum.

Proposition 3.7. *Let $S = M_{a,b} + M_{c,d} \in \mathcal{E}\ell(\mathcal{A})$ and suppose that $\sigma(S(x))$ is finite for every $x \in \mathcal{A}$. Then, for each primitive ideal P of $\mathcal{A}$, there exists $\beta_P \in \mathbb{C}$ such that*

$$\pi_P((b + \beta_P d)a),\ \pi_P(d(c - \beta_P a)) \in \mathscr{F}(X_P)$$

and either $\pi_P(b + \beta_P d)\pi_P(c - \beta_P a)$ has finite rank or $\beta_P = 0$ and $\pi_P(da)$ has finite rank.

Proof. Let $P \in \operatorname{Prim}\mathcal{A}$ and $\pi = \pi_P$, $X = X_P$. By [8, Lemma 2.1], there exists $n \in \mathbb{N}$ such that $\#\sigma(S(x)) < n$ for every $x \in \mathcal{A}$ and it therefore follows from Proposition 3.4 that there exists $\beta \in \mathbb{C}$ such that

$$\pi((b + \beta d)a),\ \pi(d(c - \beta a)) \in \mathscr{F}'(X)$$

and either $\pi(b + \beta d)\pi(c - \beta a)$ has finite rank or $\beta = 0$ and $\pi(da)$ has finite rank. Write

$$\pi(ba + \beta da) = \lambda I + F, \quad \text{where } F \in \mathscr{F}(X).$$

Suppose that $\lambda \neq 0$ for every λ associated to P. In the case that $\pi(ba) \notin \mathscr{F}'(X)$, choose vectors $\zeta_1, \dots, \zeta_r \in X$, $r \in \mathbb{N}$ such that $\{\zeta_1, \dots, \zeta_r, \pi(ba)\zeta_1, \dots, \pi(ba)\zeta_r\}$ is linearly independent modulo FX and pick $x \in \mathcal{A}$ such that

$$\pi(xba)\zeta_j = j\zeta_j,$$
$$\pi(x(\lambda - ba))\zeta_j) = 0$$

and

$$\pi(x)F = 0.$$

Then $\pi(S(x))\pi(a)\zeta_j = j\pi(a)\zeta_j$ which implies that $\{1, \dots, r\} \subseteq \sigma(S(x))$. By choosing r sufficiently large, we obtain a contradiction.

Next assume that $\pi(ba) \in \mathscr{F}'(X) \backslash \mathscr{F}(X)$. Choose vectors $\zeta_1, \dots, \zeta_r \in X$ such that $\{\pi(ba)\zeta_1, \dots, \pi(ba)\zeta_r\}$ is linearly independent. Observe that either $\pi(da) \in \mathscr{F}(X)$ or $\pi(da) \notin \mathscr{F}'(X)$. Therefore we may suppose without loss of generality that there exists $x \in \mathcal{A}$ such that $\pi(xba)\zeta_j = j\zeta_j$ and $\pi(xda)\zeta_j = 0$. Once again we get a contradiction. Now suppose that $\pi(ba) \in \mathscr{F}(X)$. Then we must have $\pi(da), \pi(dc)$ in $\mathscr{F}'(X)$. By considering $(S(x)a)\zeta_j, (S(x)c)\zeta_j$ for suitable $x \in \mathcal{A}$ and $\zeta_1, \dots, \zeta_r \in X$, we show that $\pi(dc) \in \mathscr{F}(X)$ and either $\pi(bc) \in \mathscr{F}(X)$ or $\pi(da) \in \mathscr{F}(X)$. Thus there exists $\beta_0 \in \mathbb{C}$, associated to P, such that $\pi(ba + \beta_0 da) \in \mathscr{F}(X)$. Write $S = M_{a,b+\beta_0 d} + M_{c-\beta_0 a,d}$. We already know that $\pi(dc - \beta_0 da) \in \mathscr{F}'(X)$ and $\pi((b + \beta_0 d)(c - \beta_0 a))$ has finite rank. Assume that $\pi(dc - \beta_0 da) \notin \mathscr{F}(X)$. Then, for each r, there are vectors $\zeta_1, \dots, \zeta_r \in X$ such that the set $\{\pi(dc-\beta_0 da)\zeta_1, \dots, \pi(dc-\beta_0 da)\zeta_r\}$ is linearly independent modulo $\pi((b + \beta_0 d)(c - \beta_0 a))X$. Choose $x \in \mathcal{A}$ such that $\pi(x(dc - \beta_0 da))\zeta_j = j\zeta_j$ and $\pi(x(b + \beta_0 d)(c - \beta_0 a)) = 0$. By considering $S(x)(c - \beta_0 a)$ and r sufficiently large, we get a contradiction. This completes the proof. □

For a semiprime algebra $\mathcal{A}$, we denote by $\operatorname{soc}\mathcal{A}$ its socle, see [4, Chapter IV]. As usual, $\operatorname{kh}\operatorname{soc}\mathcal{A}$ denotes the intersection of all primitive ideals of A containing soc$\mathcal{A}$, the kernel of the hull of soc$\mathcal{A}$. It is easy to show that $\mathrm{rad}(\mathcal{A}/\mathrm{soc}\mathcal{A}) = \mathrm{kh}\,\mathrm{soc}\mathcal{A}/\mathrm{soc}\mathcal{A}$.

Theorem 3.8. *Let $\mathcal{A}$ be a unital Banach algebra, and let $S = M_{a,b} + M_{c,d} \in \mathcal{E}\ell(\mathcal{A})$. If S is n-almost spectrally bounded for some $n \in \mathbb{N}$, then, for all but finitely many primitive ideals P of $\mathcal{A}$, the induced mapping $S_P\colon \mathcal{A}/P \to \mathcal{A}/P$ is spectrally bounded. Moreover, there exists $s \in \mathcal{A}$ such that $s + \mathrm{rad}\mathcal{A} \in \mathrm{soc}(\mathcal{A}/\mathrm{rad}\mathcal{A})$ and $ba + dc - s \in \mathcal{Z}(\mathcal{A})$.*

Proof. Suppose there exist distinct primitive ideals $P_1, \ldots, P_r$ such that, for every $\gamma \in \mathbb{C}$, $\pi_{P_j}((b+\gamma d)a) \notin \mathbb{C}I$. To simplify the notation we shall write X_j, π_j instead of X_{P_j}, π_{P_j} whenever convenient in the following. Fix $1 \le j \le r$. By Proposition 3.4, there exists $\beta_j \in \mathbb{C}$ such that $\pi_j((b + \beta_j d)a) \in \mathscr{F}'(X_P)$. Let $\zeta_j \in X_j$ be such that

$$\{\pi_j(ba + \beta_j da)\zeta_j, \zeta_j\}$$

is linearly independent. Choose $x_j \in \mathcal{A}$ and, for every $k \in \mathbb{N}$, choose $x_k^j \in \mathcal{A}$ such that span $\{\zeta_j, \pi_j(ba)\zeta_j, \pi_j(da)\zeta_j\}$ is invariant under $\pi_j(x_k^j)$,

$$\pi_j(\exp(x_k^j)(ba + \beta_j da))\zeta_j = \zeta_j, \qquad \pi_j(\exp(x_k^j))\zeta_j = \frac{1}{jk}\pi_j(ba + \beta_j da)\zeta_j,$$
$$\pi_j(x_j)(ba + \beta_j da)\zeta_j = 0, \qquad \pi_j(x_j)\zeta_j = \pi_j(ba + \beta_j da)\zeta_j.$$

Then $\pi\big(\exp(x_k^j)x_j \exp(-x_k^j)\big)\zeta_j = 0$. Observe that either $\pi_j(da) \in \mathbb{C}I$ or there exists $\zeta_j' \in X_j$ such that $\{\zeta_j', \pi_j(da)\zeta_j', \pi_j(ba + \beta_j da)\zeta_j'\}$ is linearly independent. In both cases, we can suppose without loss of generality that

$$\pi_j\big(\exp(x_k^j)x_j \exp(-x_k^j)da\big)\zeta_j = 0.$$

Applying the Extended Jacobson Density Theorem [13, p. 283], we can find $x \in \mathcal{A}$ and, for each k, $x_k \in \mathcal{A}$ such that

$$\pi_j(x_k)\zeta = \pi_j(x_k^j)\zeta, \quad \pi_j(x)\zeta = \pi_j(x_j)\zeta$$

for every $\zeta \in$ span $\{\zeta_j, \pi_j(ba)\zeta_j, \pi_j(da)\zeta_j\}$. Set $x_k' = \exp(x_k)x\exp(-x_k)$; then

$$\pi_j\big((Sx_k')a\big)\zeta_j = jk\pi_j(a)\zeta_j.$$

Thus $\{k, \ldots, rk\} \subseteq \sigma(Sx_k')$. As in the proof of Proposition 3.4, we show that r must be sufficiently small. We have thereby shown that for all but finitely many primitive ideals P of A, there exists $\gamma_P \in \mathbb{C}$ such that $\pi_P((b + \gamma_P d)a) \in \mathbb{C}I$. Analogously, we show that for all but finitely many primitive ideals P of $\mathcal{A}$, there exists $\gamma_P' \in \mathbb{C}$ such that $\pi_P(d(c - \gamma_P' a)) \in \mathbb{C}I$. Let Γ be the set of primitive ideals P for which there exists $\beta_P \in \mathbb{C}$ with the property that $\pi_P((b+\beta_P d)c) \in \mathscr{F}'(X_P)$. Using similar arguments, we show that for all but finitely many primitive ideals $P \in \Gamma$, there exists $\rho_P \in \mathbb{C}$ such that $\pi_P(bc + \rho_P dc) \in \mathbb{C}I$. Now let $r > n$ and let $P_1, \ldots, P_r$ be distinct primitive ideals in Γ. Suppose that for each $1 \le j \le r$, we can find $\gamma_j, \gamma_j', \rho_j \in \mathbb{C}$ and $\lambda_j, \lambda_j', \mu_j \in \mathbb{C}$ such that $\pi_j((b + \gamma_j d)a) = \lambda_j I$, $\pi_j(d(c - \gamma_j' a)) = \lambda_j' I$ and $\pi_j((b + \rho_j d)c) = \mu_j I$. In the case that $\pi_j(da) \in \mathbb{C}I$, we have $\pi_j(ba)$, $\pi_j(dc)$ and $\pi_j(bc)$ belong to $\mathbb{C}I$ so that there exists $\beta_j \in \mathbb{C}$ with $\pi_j((b+\beta_j d)(c-\beta_j a)) = 0$.

Suppose now that, for each j, there exists $\zeta_j \in X_j$ such that $\{\zeta_j, \pi_j(da)\zeta_j\}$ is linearly independent. Fix $1 \le j \le r$ and $\alpha_1, \dots, \alpha_r \in \mathbb{C}$. For each k, choose $x_k^j \in \mathcal{A}$ such that

$$\operatorname{span}\{\zeta_j, \pi_j(da)\zeta_j\}$$

is invariant under $\pi_j(x_k^j)$,

$$\pi_j\big(\exp(x_k^j)\big)\zeta_j = \frac{1}{\alpha_j k}\pi_j(da)\zeta_j \quad \text{and} \quad \pi_j\big(\exp(x_k^j)da\big)\zeta_j = \zeta_j.$$

Choose $x_j \in \mathcal{A}$ with the property that

$$\pi_j(x_j)\zeta_j = \pi_j(da)\zeta_j \quad \text{and} \quad \pi_j(x_j da)\zeta_j = \pi_j(da)\zeta_j.$$

The corresponding matrix representation of $\pi_j\big(S(\exp(x_k^j)x_j \exp(-x_k^j))\big)$ with respect to $\{\pi_j(a)\zeta_j, \pi_j(c)\zeta_j\}$ is

$$B_j = \begin{pmatrix} \lambda_j - \gamma_j\alpha_j k & \mu_j - \lambda_j'\rho_j - \alpha_j k\gamma_j'\rho_j \\ \alpha_j k & \lambda_j' + \alpha_j k\gamma_j' \end{pmatrix}.$$

Applying again the Extended Jacobson Density Theorem and arguing as above, we can find $x_k' \in \mathcal{A}$ such that $r(x_k') = r(x_k)$ and B_j is the corresponding matrix representation of $\pi_j\big(S(x_k')\big)$ with respect to $\{\pi_j(a)\zeta_j, \pi_j(c)\zeta_j\}$. The trace of B_j is $\lambda_j + \lambda_j' + \alpha_j k(\gamma_j' - \gamma_j)$. Suppose that k is sufficiently large. Using suitable scalars α_j, we see that $\gamma_j = \gamma_j'$ for at least $r-n$ primitive ideals from $\{P_1, \dots, P_r\}$. The characteristic polynomial of B_j is

$$\chi_j(t) = t^2 - (\lambda_j + \lambda_j')t - (\gamma_j^2 - \gamma_j\rho_j)(jk)^2 - (\lambda_j'\gamma_j - \lambda_j\gamma_j + \mu_j - \lambda_j'\rho_j)jk + \lambda_j\lambda_j'.$$

As above, we see that $\chi_j(t)$ has to be independent of k for at least $r-n$ primitive ideals from $\{P_1, \dots, P_r\}$. Of course, $\chi_j(t)$ is independent of k if and only if $\gamma_j = \rho_j$, and $\mu_j = \lambda_j\gamma_j$ or $\gamma_j = 0$ and $\mu_j = \lambda_j'\rho_j$. Thus, for all but finitely many $P \in \Gamma$, there exists $\beta_P \in \mathbb{C}$ such that $\pi_P((b+\beta_P d)a)$, $\pi_P(d(c-\beta_P a))$ belong to $\mathbb{C}I$ and $\pi_P(b+\beta_P d)\pi_P(c-\beta_P a) = 0$. Next let Γ' be the set of primitive ideals for which $\pi_P(ba), \pi_P(dc), \pi_P(da)$ belong to $\mathbb{C}I$. In the same manner we see that for all but finitely many $P \in \Gamma' \setminus \Gamma$ we have $\pi_P(da) = 0$.

In summary we proved that for all but finitely many primitive ideals P of A, there exists $\beta_P \in \mathbb{C}$ such that

$$\pi_P((b+\beta_P d)a) \in \mathbb{C}I \quad \text{and} \quad \pi_P(d(c-\beta_P a)) \in \mathbb{C}I \tag{3.4}$$

and either $\pi_P(b+\beta_P d)\pi_P(c-\beta_P a) = 0$ or $\beta_P = 0$ and $\pi_P(da) = 0$. By Proposition 2.3, the induced map S_P is spectrally bounded for all but finitely many primitive ideals P of $\mathcal{A}$.

By (3.4), for all but finitely many $P \in \operatorname{Prim}\mathcal{A}$, $\big[\pi_P(ba+dc), \pi_P(x)\big] = 0$ for all $x \in \mathcal{A}$. For the remaining primitive ideals Q, $\pi_Q(ba+dc) \in \mathscr{F}'(X_Q)$ by Proposition 3.4 and hence $\big[\pi_Q(ba+dc), \pi_Q(x)\big]$ has finite spectrum for all $x \in \mathcal{A}$. It follows from (2.4) that $\#\sigma([ba+dc, x]) < \infty$ for all $x \in \mathcal{A}$. Using [6,

Theorem 2.4], we find that there exists $s \in \mathcal{A}$ such that $s + \mathrm{rad}\mathcal{A} \in \mathrm{soc}(\mathcal{A}/\mathrm{rad}\mathcal{A})$ and $ba + dc - s \in \mathcal{Z}(\mathcal{A})$. This completes the proof. □

The following theorem is our second main result.

Theorem 3.9. *Let $\mathcal{A}$ be a semisimple unital Banach algebra, and let $S = M_{a,b} + M_{c,d} \in \mathcal{E}\ell(\mathcal{A})$. If every element in the range of S has finite spectrum then $S(x)^3 \in \mathrm{soc}\mathcal{A}$ for every $x \in \mathcal{A}$.*

Proof. Let $P \in \mathrm{Prim}\mathcal{A}$. Using the notation of Proposition 3.7 we write S_P in the form

$$S_P = M_{a_P, b_P + \beta_P d_P} + M_{c_P - \beta_P a_P, d_P}. \tag{3.5}$$

By the statement of Proposition 3.7, we check at once that, for each P, $S_P(x_P)^3 \in \mathrm{soc}(\mathcal{A}_P)$ for all $x \in \mathcal{A}$.

By the proof of Theorem 3.8, for all but finitely many primitive ideals P, $b_P a_P + \beta_P d_P a_P = 0$, $d_P c_P + b_P a_P = 0$ and either $(b_P + \beta_P d_P)(c_P - \beta_P a_P) = 0$ or $\beta_P = 0$ and $d_P a_P = 0$. Invoking (3.5) we find that $S_P(x_P)^3 = 0$ in this case. Applying, e.g., [8, Proposition 2.2], we infer that $(Sx)^3 \in \mathrm{soc}\mathcal{A}$ for all $x \in \mathcal{A}$, as desired. □

We note some consequences for length one elementary operators.

Corollary 3.10. *Let $M_{a,b} \in \mathcal{E}\ell(\mathcal{A})$ for a semisimple unital Banach algebra $\mathcal{A}$. Then every element in the range of $M_{a,b}$ has finite spectrum if and only if $ba \in \mathrm{soc}\mathcal{A}$. In this case $M_{a,b}(x)^2 \in \mathrm{soc}\mathcal{A}$ for all $x \in \mathcal{A}$.*

Proof. Since $\sigma(axb) \cup \{0\} = \sigma(bax) \cup \{0\}$ for each $x \in \mathcal{A}$, the "if"-part is evident. For the "only if"-part we specialise Proposition 3.7 and the proof of Theorem 3.8 to the case at hand to obtain that $b_P a_P \in \mathrm{soc}(\mathcal{A}_P)$ for every $P \in \mathrm{Prim}\mathcal{A}$ and $b_P a_P = 0$ except for at most finitely many P's. The characterisation of the socle given in [8, Proposition 2.2] allows us to conclude that $ba \in \mathrm{soc}\mathcal{A}$. □

There is an obvious extension of Corollary 3.10 to general Banach algebras.

Corollary 3.11. *Let $M_{a,b} \in \mathcal{E}\ell(\mathcal{A})$ be n-almost spectrally bounded for some $n \in \mathbb{N}$. Then there exists $s \in \mathcal{A}$ such that $s + \mathrm{rad}\mathcal{A} \in \mathrm{soc}(\mathcal{A}/\mathrm{rad}\mathcal{A})$ and $ba - s \in \mathcal{Z}(\mathcal{A})$.*

Proof. This is immediate from Theorem 3.8. □

Suppose that $\mathcal{A}$ is semisimple and let $T\colon \mathcal{A} \to \mathcal{A}$ be a linear operator. We say that T is essentially spectrally bounded if there exists $M > 0$ such that

$$r(Tx + I(\mathcal{A})) \le M r(x + I(\mathcal{A})) \qquad (x \in \mathcal{A}),$$

where $I(\mathcal{A}) = \mathrm{kh}\,\mathrm{soc}\mathcal{A}$ is the ideal of inessential elements in $\mathcal{A}$ (see [15] for this notion in a more general context).

Corollary 3.12. *Let $M_{a,b} \in \mathcal{E}\ell(\mathcal{A})$ be n-almost spectrally bounded for some $n \in \mathbb{N}$, where $\mathcal{A}$ is semisimple. Then $M_{a,b}$ is essentially spectrally bounded.*

Proof. Denote by $\hat{\mathcal{A}} = \mathcal{A}/I(A)$ the quotient Banach algebra and by $M_{\hat{a},\hat{b}}$ the induced elementary operator on $\hat{\mathcal{A}}$. If $M_{a,b}$ is n-almost spectrally bounded, then $\hat{b}\hat{a} = ba + I(\mathcal{A}) \in Z(\hat{\mathcal{A}})$ by Corollary 3.11. Consequently, $M_{\hat{a},\hat{b}}$ is spectrally bounded which is the same as the assertion. □

We do not know if the same result holds for elementary operators of length two.

Concluding Remark 3.13. In view of the above results it is tempting to surmise that, whenever $\mathcal{A}$ is semisimple and S is a spectrally bounded elementary operator of length two, there exist $a, b, c, d \in \mathcal{A}$ such that $S = M_{a,b} + M_{c,d}$ *and* both $M_{a,b}$ and $M_{c,d}$ are spectrally bounded with orthogonal ranges. However, the centre of $\mathcal{A}$, and that of a suitable Banach algebra enlargement of $\mathcal{A}$, may not be well behaved enough to allow for this.

References

[1] S.A. Amitsur, *Generalized polynomial identities and pivotal monomials*, Trans. Amer. Math. Soc. **114** (1965), 210–226.

[2] P. Ara and M. Mathieu, *Local multipliers of C*-algebras*, Springer Monographs in Mathematics, Springer-Verlag, London, 2003.

[3] B. Aupetit, *A primer on spectral theory*, Springer-Verlag, New York, 1991.

[4] F.F. Bonsall and J. Duncan, *Complete normed algebras*, Springer-Verlag, Berlin, New York, 1973.

[5] N. Boudi, *On the product of derivations in Banach algebras*, Math. Proc. Royal Irish Academy **109 A** (2009), 201–211.

[6] N. Boudi and M. Mathieu, *Commutators with finite spectrum*, Illinois J. Math. **48** (2004), 687–699.

[7] M. Brešar, *Derivations decreasing the spectral radius*, Arch. Math. **61** (1993), 160–162.

[8] M. Brešar and P. Šemrl, *Derivations mapping into the socle*, Math. Proc. Cambridge Phil. Soc. **120** (1996), 339–346.

[9] C. Costara and D. Repovš, *Spectral isometries onto algebras having a separating family of finite-dimensional irreducible representations*, J. Math. Anal. Appl. **365** (2010), 605–608.

[10] J. Cui and J. Hou, *The spectrally bounded linear maps on operator algebras*, Studia Math. **150** (2002), 261–271.

[11] A. Fošner and P. Šemrl, *Spectrally bounded linear maps on $\mathcal{B}(X)$*, Canad. Math. Bull. **47** (2004), 369–372.

[12] R. Curto and M. Mathieu, *Spectrally bounded generalized inner derivations*, Proc. Amer. Math. Soc. **123** (1995), 2431–2434.

[13] J.M.G. Fell and R.S. Doran, *Representations of *-algebras, locally compact groups, and Banach *-algebraic bundles*, I, Academic Press, New York, 1988.

[14] M. Mathieu, *Where to find the image of a derivation*, Banach Center Publ. **30** (1994), 237–249.

[15] M. Mathieu, *A collection of problems on spectrally bounded operators*, Asian-Eur. J. Math. **2** (2009), 487–501.

[16] M. Mathieu and G.J. Schick, *First results on spectrally bounded operators*, Studia Math. **152** (2002), 187–199.

[17] M. Mathieu and G.J. Schick, *Spectrally bounded operators from von Neumann algebras*, J. Operator Theory **49** (2003), 285–293.

[18] M. Mathieu and A.R. Sourour, *Hereditary properties of spectral isometries*, Arch. Math. **82** (2004), 222–229.

[19] A. Moreno Galindo and A. Rodríguez Palacios, *Topologically nilpotent normed algebras*, preprint 2010.

[20] V. Pták, *Derivations, commutators and the radical*, Manuscripta Math. **23** (1978), 355–362.

Nadia Boudi
Département de Mathématiques
Université Moulay Ismail
Faculté des Sciences
Meknes, Maroc
e-mail: nadia_boudi@hotmail.com

Martin Mathieu
Department of Pure Mathematics
Queen's University Belfast
Belfast BT7 1NN
Northern Ireland
e-mail: m.m@qub.ac.uk

The Browder Spectrum of an Elementary Operator

Derek Kitson

Abstract. We relate the ascent and descent of n-tuples of multiplication operators $M_{a,b}(u) = aub$ to that of the coefficient Hilbert space operators a, b. For example, if $\mathbf{a} = (a_1, \dots, a_n)$ and $\mathbf{b}^* = (b_1^*, \dots, b_m^*)$ have finite non-zero ascent and descent s and t, respectively, then the $(n+m)$-tuple $(L_{\mathbf{a}}, R_{\mathbf{b}})$ of left and right multiplication operators has finite ascent and descent $s+t-1$. Using these results we obtain a description of the Browder joint spectrum of $(L_{\mathbf{a}}, R_{\mathbf{b}})$ and provide formulae for the Browder spectrum of an elementary operator acting on $B(H)$ or on a norm ideal of $B(H)$.

Mathematics Subject Classification (2000). Primary 47B49; Secondary 47A13.

Keywords. Ascent, descent, joint spectrum.

Introduction

The Browder spectrum of a Banach space operator is obtained by removing from the ordinary spectrum all eigenvalues of finite multiplicity which are poles of the resolvent. Let H be a complex Hilbert space and $B(H)$ the collection of bounded operators on H. An elementary operator on $B(H)$ is an operator of the form $\mathcal{E} : B(H) \to B(H)$, $u \mapsto a_1ub_1 + \cdots + a_nub_n$ where $\mathbf{a} = (a_1, \dots, a_n)$, $\mathbf{b} = (b_1, \dots, b_n) \in B(H)^n$. Spectral properties of elementary operators have been considered by various authors (see [2, 3, 5, 6, 7, 9, 10]). In this article we obtain formulae for the Browder spectrum of $\mathcal{E}$. These formulae may have applications to Weyl and Browder type theorems for elementary operators.

The ascent of a linear mapping a acting on a vector space X is usually described as the length of the increasing chain of null spaces

$$\{0\} \subseteq \ker a \subseteq \ker a^2 \subseteq \ker a^3 \subseteq \cdots$$

R.E. Curto and M. Mathieu (eds.), *Elementary Operators and Their Applications*, Operator Theory: Advances and Applications 212, DOI 10.1007/978-3-0348-0037-2_2, © Springer Basel AG 2011

This value is also the smallest non-negative integer r for which $\ker a \cap \operatorname{ran} a^r = \{0\}$. The descent of a is the length of the decreasing chain of range spaces

$$X \supseteq \operatorname{ran} a \supseteq \operatorname{ran} a^2 \supseteq \operatorname{ran} a^3 \supseteq \cdots$$

and equals the smallest r for which $\ker a^r + \operatorname{ran} a = X$. Let $\mathbf{a} = (a_1, \dots, a_n)$ be an n-tuple of linear mappings on a vector space X. We write $N(\mathbf{a}) = \bigcap_{j=1}^n \ker a_j$ and $R(\mathbf{a}) = \sum_{j=1}^n \operatorname{ran} a_j$. We denote by $\mathbf{a}^r$ the lexicographically ordered n^r-tuple of operators on X consisting of all products $a_{j_1} \dots a_{j_r}$. For example if $n = 2$ then

$$\begin{aligned} \mathbf{a} &= (a_1, a_2) \\ \mathbf{a}^2 &= (a_1^2,\ a_1a_2,\ a_2a_1,\ a_2^2) \\ \mathbf{a}^3 &= (a_1^3,\ a_1^2a_2,\ a_1a_2a_1,\ a_1a_2^2,\ a_2a_1^2,\ a_2a_1a_2,\ a_2^2a_1,\ a_2^3) \quad \text{etc.} \end{aligned}$$

In the terminology of [8], the ascent of $\mathbf{a} = (a_1, \dots, a_n)$ on X is the smallest r such that $N(\mathbf{a}) \cap R(\mathbf{a}^r) = \{0\}$. The descent of $\mathbf{a} = (a_1, \dots, a_n)$ on X is the smallest r such that $N(\mathbf{a}^r) + R(\mathbf{a}) = X$. If the ascent and descent of $\mathbf{a} = (a_1, \dots, a_n)$ are both finite then they are equal ([8, Proposition 1.9]). We will denote the ascent of an n-tuple $\mathbf{a} = (a_1, \dots, a_n)$ by $\alpha(\mathbf{a}, X)$ or simply $\alpha(\mathbf{a})$ if the space X is understood. Similarly we write $\delta(\mathbf{a}, X)$ or simply $\delta(\mathbf{a})$ for the descent of $\mathbf{a} = (a_1, \dots, a_n)$.

For operators $a, b \in B(H)$ define the multiplication operator $M_{a,b} : B(H) \to B(H)$, $u \mapsto aub$. With I the identity operator on H, the left multiplication operator is $L_a = M_{a,I}$ and the right multiplication operator is $R_b = M_{I,b}$. Given an n-tuple $\mathbf{a} = (a_1, \dots, a_n)$ of operators on H we write $L_{\mathbf{a}} = (L_{a_1}, \dots, L_{a_n})$ and $R_{\mathbf{a}} = (R_{a_1}, \dots, R_{a_n})$. More generally, we will consider multiplication operators acting on ideals $\mathcal{I}$ of $B(H)$ such as the compact operators, the trace-class operators, the Hilbert-Schmidt operators and all p-Schatten classes of operators.

In Section 1 we show that if $\mathbf{a} = (a_1, \dots, a_n)$ and $\mathbf{b}^* = (b_1^*, \dots, b_m^*)$ have finite non-zero ascent and descent s and t respectively then the $(n+m)$-tuple $(L_{\mathbf{a}}, R_{\mathbf{b}}) = (L_{a_1}, \dots, L_{a_n}, R_{b_1}, \dots, R_{b_m})$ has finite ascent and descent $s + t - 1$. We also obtain bounds for the ascent and descent of $M_{\mathbf{a},\mathbf{b}} = (M_{a_i,b_j})_{i=1,j=1}^{n,\,m}$. In [8] a Browder joint spectrum σ_b was introduced for n-tuples of Banach space operators. In Section 2 we obtain a description of the Browder joint spectrum of $(L_{\mathbf{a}}, R_{\mathbf{b}})$ and collect formulae for the Browder spectrum of an elementary operator acting on $B(H)$ or on a norm ideal of $B(H)$,

$$\begin{aligned} \sigma_b(\mathcal{E}) &= \big(\sigma_l(\mathbf{a}) \circ \sigma_b^-(\mathbf{b})\big) \cup \big(\sigma_b^+(\mathbf{a}) \circ \sigma_r(\mathbf{b})\big) \cup \big(\sigma_r(\mathbf{a}) \circ \sigma_b^+(\mathbf{b})\big) \cup \big(\sigma_b^-(\mathbf{a}) \circ \sigma_l(\mathbf{b})\big) \\ &= (\sigma_H(\mathbf{a}) \circ \sigma_b(\mathbf{b})) \cup (\sigma_b(\mathbf{a}) \circ \sigma_H(\mathbf{b})) \\ &= (\sigma_T(\mathbf{a}) \circ \sigma_{Tb}(\mathbf{b})) \cup (\sigma_{Tb}(\mathbf{a}) \circ \sigma_T(\mathbf{b})) \end{aligned}$$

Here σ_l and σ_r are the left and right spectra, σ_b^+ and σ_b^- are analogues of the semi-Browder spectra of an operator, σ_H is the Harte spectrum, σ_T is the Taylor spectrum and σ_{Tb} is the Taylor–Browder spectrum.

1. Ascent and descent

Theorem 1.1. *Let* $\mathbf{a} = (a_1, \ldots, a_n)$ *be an* n*-tuple of operators on a Hilbert space* H. *Let* $\mathcal{I}$ *be a left ideal of* $B(H)$. *Then*

(i) $\alpha(L_{\mathbf{a}}, \mathcal{I}) = \alpha(\mathbf{a}, H)$
(ii) $\delta(L_{\mathbf{a}}, \mathcal{I}) = \delta(\mathbf{a}, H)$

Proof. (i) If $u \in N(L_{\mathbf{a}}) \cap R((L_{\mathbf{a}})^r)$ then $\operatorname{ran} u \subseteq N(\mathbf{a}) \cap R(\mathbf{a}^r)$. We conclude that $\alpha(L_{\mathbf{a}}) \leq \alpha(\mathbf{a})$. For the reverse inequality, suppose $x \in N(\mathbf{a}) \cap R(\mathbf{a}^r)$. Then $x = \sum a_{i_1} \ldots a_{i_r}(x_{i_1 \ldots i_r})$ for some $x_{i_1 \ldots i_r} \in H$. Choose $y \in H$ such that $\langle x, y\rangle = 1$. For each $x_{i_1 \ldots i_r}$ consider the rank one operator $p_{i_1 \ldots i_r} \in \mathcal{I}$ given by $p_{i_1 \ldots i_r}(z) = \langle z, y\rangle x_{i_1 \ldots i_r}$ for all $z \in H$. Let $p = \sum a_{i_1} \ldots a_{i_r} p_{i_1 \ldots i_r}$. Then $p \in N(L_{\mathbf{a}}) \cap R((L_{\mathbf{a}})^r)$ and $x = p(x)$. We conclude that $\alpha(\mathbf{a}) \leq \alpha(L_{\mathbf{a}})$.

(ii) Suppose $N((L_{\mathbf{a}})^r) + R(L_{\mathbf{a}}) = \mathcal{I}$. Let $x \in H$ and let $p \in \mathcal{I}$ be a rank one operator with $p(x) = x$. Then $p = u + v$ for some $u \in N((L_{\mathbf{a}})^r)$ and some $v \in R(L_{\mathbf{a}})$. We have $v = L_{a_1}(v_1) + \cdots + L_{a_n}(v_n)$ for some $v_1, \ldots, v_n \in \mathcal{I}$. Now $x = (u+v)(x) = u(x) + (a_1 v_1 + \cdots + a_n v_n)(x) \in N(\mathbf{a}^r) + R(\mathbf{a})$. Hence $N(\mathbf{a}^r) + R(\mathbf{a}) = H$ and so $\delta(\mathbf{a}) \leq \delta(L_{\mathbf{a}})$.

For the reverse inequality, suppose $N(\mathbf{a}^r) + R(\mathbf{a}) = H$. Let $q \in B(H)$ be the orthogonal projection onto $N(\mathbf{a}^r)$. Define $T : H^{n+1} \to H$ by $T(x, y_1, \ldots, y_n) = q(x) + a_1(y_1) + \cdots + a_n(y_n)$. Note that T is surjective and so there exists a right inverse operator $C : H \to H^{n+1}$. Write $C(z) = (f(z), g_1(z), \ldots, g_n(z))$ where $f, g_1, \ldots, g_n \in B(H)$. Then for all $u \in \mathcal{I}$ we have

$$u = (T \circ C)u = (q \circ f)u + (a_1 \circ g_1)u + \cdots + (a_n \circ g_n)u \in N((L_{\mathbf{a}})^r) + R(L_{\mathbf{a}})$$

We conclude that $N((L_{\mathbf{a}})^r) + R(L_{\mathbf{a}}) = \mathcal{I}$ and so $\delta(L_{\mathbf{a}}) \leq \delta(\mathbf{a})$. □

By appealing to adjoints we can obtain further equalities. Given an n-tuple $\mathbf{a} = (a_1, \ldots, a_n)$ we denote by $\mathbf{a}^* = (a_1^*, \ldots, a_n^*)$ the n-tuple of adjoint operators.

Theorem 1.2. *Let* $\mathbf{a} = (a_1, \ldots, a_n)$ *be an* n*-tuple of operators on a Hilbert space* H. *Let* $\mathcal{I}$ *be a two-sided ideal of* $B(H)$. *Then*

(i) $\alpha(R_{\mathbf{a}}, \mathcal{I}) = \alpha(\mathbf{a}^*, H)$
(ii) $\delta(R_{\mathbf{a}}, \mathcal{I}) = \delta(\mathbf{a}^*, H)$

Proof. Note that the operator $R_a : \mathcal{I} \to \mathcal{I}$ is similar to the operator $L_{a^*} : \mathcal{I} \to \mathcal{I}$ where similarity is provided by the involution $\tau : \mathcal{I} \to \mathcal{I}$, $b \mapsto b^*$. Applying this similarity we have $\alpha(R_{\mathbf{a}}, \mathcal{I}) = \alpha(L_{\mathbf{a}^*}, \mathcal{I})$ and $\delta(R_{\mathbf{a}}, \mathcal{I}) = \delta(L_{\mathbf{a}^*}, \mathcal{I})$. The result now follows from Theorem 1.1. □

Remark 1.3. The ascent and descent of the right multiplication R_a behave somewhat differently to left multiplication. For example if a is the unilateral shift operator on ℓ^2 then $\delta(a, \ell^2) = \delta(L_a, B(\ell^2)) = \infty$ but $\delta(R_a, B(\ell^2)) = 0$. (This is noted in [1, Remark 2.3].) For a general Hilbert space operator, if a^{r+1} has closed range where $r = \alpha(a)$ then $\alpha(a, H) = \delta(R_a, B(H))$. The necessity of the closed range condition is illustrated by the self-adjoint operator $a(e_j) = \frac{1}{2^j} e_j$ on ℓ^2. In this case

$\alpha(a,\ell^2) = 0$ and a does not have closed range. It follows from the above theorem that $\delta(R_a, B(\ell^2)) = \infty$.

For a single operator $a \in B(H)$ with finite ascent and descent we have equalities $\alpha(a) = \delta(a) = \alpha(a^*) = \delta(a^*) = \alpha(L_a) = \delta(L_a) = \alpha(R_a) = \delta(R_a)$. The following example shows that this is not always true in the multivariable case.

Example. Let H be the Hilbert space with orthonormal basis $(e_{i,j})_{i,j=1}^{\infty}$. Define $\mathbf{a} = (a_1, a_2)$ where

$$a_1(e_{i,j}) = \begin{cases} e_{i-1,j} & \text{if } i > 1 \\ 0 & \text{if } i = 1 \end{cases} \quad \text{and} \quad a_2(e_{i,j}) = \frac{1}{2^j} e_{i,j}$$

Then $\alpha(\mathbf{a}) = \delta(\mathbf{a}) = 0$ but $\alpha(\mathbf{a}^*) = 0$ and $\delta(\mathbf{a}^*) = \infty$.

Theorem 1.4. *Let $\mathbf{a} = (a_1, \dots, a_n)$ and $\mathbf{b} = (b_1, \dots, b_m)$ be tuples of operators on a Hilbert space H. Let $\mathcal{I}$ be a two-sided ideal of $B(H)$.*

(i) *If $\alpha(\mathbf{a}) > 0$ and $\alpha(\mathbf{b}^*) > 0$ then $\alpha(\mathbf{a}) + \alpha(\mathbf{b}^*) - 1 \leq \alpha((L_{\mathbf{a}}, R_{\mathbf{b}}), \mathcal{I})$.*
(ii) *If $\delta(\mathbf{a}) > 0$ and $\alpha(\mathbf{b}) > 0$ then $\delta(\mathbf{a}) + \alpha(\mathbf{b}) - 1 \leq \delta((L_{\mathbf{a}}, R_{\mathbf{b}}), \mathcal{I})$.*

Proof. (i) Suppose there exist non-zero elements $x \in N(\mathbf{a}) \cap R(\mathbf{a}^s)$ and $y \in N(\mathbf{b}^*) \cap R((\mathbf{b}^*)^t)$. Define the rank one operator $p \in \mathcal{I}$ by $p(z) = \langle z, y\rangle x$ for all $z \in H$. Now $x = \sum a_{i_1} \dots a_{i_s}(x_{i_1 \dots i_s})$ for some $x_{i_1\dots i_s} \in H$. Also $y = \sum b_{j_1}^* \dots b_{j_t}^*(y_{j_1\dots j_t})$ for some $y_{j_1 \dots j_t} \in H$. For each $i_1, \dots, i_s = 1, \dots, n$ and each $j_1, \dots, j_t = 1, \dots, m$ define the rank one operator $p_{i_1 \dots i_s j_1 \dots j_t} \in \mathcal{I}$ by $p_{i_1\dots i_s j_1 \dots j_t}(z) = \langle z, y_{j_1 \dots j_t}\rangle x_{i_1 \dots i_s}$ for all $z \in H$. Then we can verify that $p = \sum a_{i_1} \dots a_{i_s} p_{i_1 \dots i_s j_1 \dots j_t} b_{j_1} \dots b_{j_t}$. We have $p \in N(L_{\mathbf{a}}, R_{\mathbf{b}}) \cap R((L_{\mathbf{a}}, R_{\mathbf{b}})^{s+t})$ and so $s+t+1 \leq \alpha(L_{\mathbf{a}}, R_{\mathbf{b}})$. It follows that $\alpha(\mathbf{a}) + \alpha(\mathbf{b}^*) - 1 \leq \alpha(L_{\mathbf{a}}, R_{\mathbf{b}})$.

(ii) Suppose there exists $x \in N(\mathbf{b}) \cap R(\mathbf{b}^t)$ with $\|x\| = 1$ and suppose there exists $y \in H$ with $y \notin N(\mathbf{a}^s) + R(\mathbf{a})$. Define the rank one operator $q \in \mathcal{I}$ by $q(z) = \langle z, x\rangle y$ for all $z \in H$. If $\delta(L_{\mathbf{a}}, R_{\mathbf{b}}) \leq s+t$ then $N((L_{\mathbf{a}}, R_{\mathbf{b}})^{s+t}) + R(L_{\mathbf{a}}, R_{\mathbf{b}}) = \mathcal{I}$. Thus we can write $q = v + w$ for some $v \in N((L_{\mathbf{a}}, R_{\mathbf{b}})^{s+t})$ and some $w \in R(L_{\mathbf{a}}, R_{\mathbf{b}})$. Now $v(x) \in N(\mathbf{a}^s)$ and $w(x) \in R(\mathbf{a})$. Thus $y = q(x) = v(x) + w(x) \in N(\mathbf{a}^s) + R(\mathbf{a})$ which is a contradiction. We conclude that $s+t+1 \leq \delta(L_{\mathbf{a}}, R_{\mathbf{b}})$. It now follows that $\delta(\mathbf{a}) + \alpha(\mathbf{b}) - 1 \leq \delta((L_{\mathbf{a}}, R_{\mathbf{b}}), \mathcal{I})$. □

Corollary 1.5. *Let $\mathbf{a} = (a_1, \dots, a_n)$ and $\mathbf{b} = (b_1, \dots, b_m)$ be tuples of operators on a Hilbert space H. Let $\mathcal{I}$ be a two-sided ideal of $B(H)$. Suppose $s = \alpha(\mathbf{a}) = \delta(\mathbf{a}) < \infty$ and $t = \alpha(\mathbf{b}^*) = \delta(\mathbf{b}^*) < \infty$. Then*

(i) $\alpha((L_{\mathbf{a}}, R_{\mathbf{b}}), \mathcal{I}) = \delta((L_{\mathbf{a}}, R_{\mathbf{b}}), \mathcal{I}) = 0$ *if $s = 0$ or $t = 0$;*
(ii) $\alpha((L_{\mathbf{a}}, R_{\mathbf{b}}), \mathcal{I}) = \delta((L_{\mathbf{a}}, R_{\mathbf{b}}), \mathcal{I}) = s+t-1$ *if $s, t > 0$.*

Proof. By Theorem 1.1 we have $\alpha(L_{\mathbf{a}}) = \delta(L_{\mathbf{a}}) = s$ and by Theorem 1.2, $\alpha(R_{\mathbf{b}}) = \delta(R_{\mathbf{b}}) = t$. (i) is clear so suppose $s, t > 0$. The operators in $L_{\mathbf{a}}$ commute with the operators in $R_{\mathbf{b}}$ and so from the argument in [8, Proposition 2.1], $(L_{\mathbf{a}}, R_{\mathbf{b}})$ has finite ascent and descent at most $s+t-1$. Now apply Theorem 1.4. □

Lemma 1.6. *Let* $\mathbf{s} = (s_1, \ldots, s_n)$ *and* $\mathbf{t} = (t_1, \ldots, t_m)$ *be tuples of linear mappings on a vector space* X *such that* $s_i t_j = t_j s_i$ *for all* i, j. *Let* $\mathbf{st} = (s_i t_j)_{i=1,j=1}^{n,m}$. *Then*

(i) $\alpha(\mathbf{st}) \leq \max(\alpha(\mathbf{s}),\, \alpha(\mathbf{t}))$
(ii) $\delta(\mathbf{st}) \leq \max(\delta(\mathbf{s}),\, \delta(\mathbf{t}))$

Proof. Suppose $r = \max(\alpha(\mathbf{s}),\, \alpha(\mathbf{t})) < \infty$. If $x \in N(\mathbf{st}) \cap R((\mathbf{st})^r)$ then $t_j(x) \in N(\mathbf{s}) \cap R(\mathbf{s}^r) = \{0\}$ for all j Hence $x \in N(\mathbf{t}) \cap R(\mathbf{t}^r) = \{0\}$. We conclude that $\alpha(\mathbf{st}) \leq r$. Considering transpose operators acting on the algebraic conjugate of X (and noting [8, Proposition 3.4]) we have $\delta(\mathbf{st}) = \alpha((\mathbf{st})') = \alpha(\mathbf{s}'\mathbf{t}') \leq \max(\alpha(\mathbf{s}'),\, \alpha(\mathbf{t}')) = \max(\delta(\mathbf{s}),\, \delta(\mathbf{t}))$. □

Theorem 1.7. *Let* $\mathbf{a} = (a_1, \ldots, a_n)$ *and* $\mathbf{b} = (b_1, \ldots, b_m)$ *be tuples of operators on a Hilbert space* H. *Let* $\mathcal{I}$ *be a two-sided ideal of* $B(H)$ *and let* $M_{\mathbf{a},\mathbf{b}} = (M_{a_i,b_j})_{i=1,j=1}^{n,\,m}$. *Then*

(i) $\min(\alpha(\mathbf{a}),\, \alpha(\mathbf{b}^*)) \leq \alpha(M_{\mathbf{a},\mathbf{b}},\, \mathcal{I}) \leq \max(\alpha(\mathbf{a}),\, \alpha(\mathbf{b}^*))$
(ii) $\min(\delta(\mathbf{a}),\, \alpha(\mathbf{b})) \leq \delta(M_{\mathbf{a},\mathbf{b}},\, \mathcal{I}) \leq \max(\delta(\mathbf{a}),\, \delta(\mathbf{b}^*))$

Proof. By Lemma 1.6 we have $\alpha(M_{\mathbf{a},\mathbf{b}}) \leq \max(\alpha(L_{\mathbf{a}}),\, \alpha(R_{\mathbf{b}}))$ and $\delta(M_{\mathbf{a},\mathbf{b}}) \leq \max(\delta(L_{\mathbf{a}}),\, \delta(R_{\mathbf{b}}))$. Now Theorem 1.1 and Theorem 1.2 show the rightmost inequalities in (i) and (ii) hold.

Suppose there are non-zero elements $x \in N(\mathbf{a}) \cap R(\mathbf{a}^s)$ and $y \in N(\mathbf{b}^*) \cap R((\mathbf{b}^*)^t)$. Define $p \in \mathcal{I}$ by $p(z) = \langle z, y\rangle x$. Then $p \in N(M_{\mathbf{a},\mathbf{b}}) \cap R((M_{\mathbf{a},\mathbf{b}})^r)$ where $r = \min(s, t)$. It follows that $\min(\alpha(\mathbf{a}),\, \alpha(\mathbf{b}^*)) \leq \alpha(M_{\mathbf{a},\mathbf{b}})$.

Suppose there exists $x \in H$ with $x \notin N(\mathbf{a}^s) + R(\mathbf{a})$ and $y \in N(\mathbf{b}) \cap R(\mathbf{b}^t)$ with $\|y\| = 1$. Define $p \in \mathcal{I}$ by $p(z) = \langle z, y\rangle x$. Let $r = \min(s, t)$ and suppose $N((M_{\mathbf{a},\mathbf{b}})^r) + R(M_{\mathbf{a},\mathbf{b}}) = \mathcal{I}$. Then $p = v + w$ for some $v \in N((M_{\mathbf{a},\mathbf{b}})^r)$ and some $w \in R(M_{\mathbf{a},\mathbf{b}})$. Now $v(y) \in N(\mathbf{a}^s)$ and $w(y) = 0$. Hence $x = p(y) = v(y) \in N(\mathbf{a}^s)$. This is a contradiction and so $\min(s, t) < \delta(M_{\mathbf{a},\mathbf{b}})$. The result follows. □

2. Browder spectrum of $(L_{\mathbf{a}},\, R_{\mathbf{b}})$

Throughout this section H is a complex Hilbert space and we consider multiplication operators acting on $B(H)$ or on a norm ideal $\mathcal{I}$ of $B(H)$. For a general reference on joint spectra we cite [11]. An n-tuple $\mathbf{a} = (a_1, \ldots, a_n)$ of operators on a Banach space X is left (resp. right) invertible if there exists $b_1, \ldots, b_n \in B(X)$ with $b_1 a_1 + \cdots + b_n a_n = I$ (resp. $a_1 b_1 + \cdots + a_n b_n = I$). Denote by I^{left} the collection of left invertible tuples and by I^{right} the collection of right invertible tuples. Recall the left and right spectra for a tuple of operators are

$$\sigma_l(\mathbf{a}) = \{\lambda \in \mathbb{C}^n : \mathbf{a} - \lambda \notin I^{\text{left}}\}$$
$$\sigma_r(\mathbf{a}) = \{\lambda \in \mathbb{C}^n : \mathbf{a} - \lambda \notin I^{\text{right}}\}$$

An n-tuple $\mathbf{a} = (a_1, \ldots, a_n)$ is called upper semi-Fredholm if the column operator $X \to X^n$, $x \mapsto (a_1(x), \ldots, a_n(x))$ has finite-dimensional kernel and closed range. An n-tuple $\mathbf{a} = (a_1, \ldots, a_n)$ is lower semi-Fredholm if the row operator $X^n \to$

X, $(x_1, \ldots, x_n) \mapsto a_1(x) + \cdots + a_n(x)$ has finite codimensional range. We denote by F^+ and F^- respectively the collection of upper and lower semi-Fredholm tuples. The upper and lower Fredholm spectra are

$$\sigma_e^+(\mathbf{a}, X) = \{\lambda \in \mathbb{C}^n : \mathbf{a} - \lambda \notin F^+\}$$
$$\sigma_e^-(\mathbf{a}, X) = \{\lambda \in \mathbb{C}^n : \mathbf{a} - \lambda \notin F^-\}$$

An n-tuple is called Fredholm if it is both upper and lower semi-Fredholm. We will make use of the following formulae which can be found in [2].

$$\sigma_e^+((L_\mathbf{a}, R_\mathbf{b}), \mathcal{I}) = \big(\sigma_l(\mathbf{a}) \times \sigma_e^-(\mathbf{b})\big) \cup \big(\sigma_e^+(\mathbf{a}) \times \sigma_r(\mathbf{b})\big) \tag{2.1}$$
$$\sigma_e^-((L_\mathbf{a}, R_\mathbf{b}), \mathcal{I}) = \big(\sigma_r(\mathbf{a}) \times \sigma_e^+(\mathbf{b})\big) \cup \big(\sigma_e^-(\mathbf{a}) \times \sigma_l(\mathbf{b})\big) \tag{2.2}$$

We denote by B^+ the collection of upper semi-Fredholm tuples with finite ascent and by B^- the collection of lower semi-Fredholm tuples with finite descent. Define

$$\sigma_b^+(\mathbf{a}) = \{\lambda \in \mathbb{C}^n : \mathbf{a} - \lambda \notin B^+\}$$
$$\sigma_b^-(\mathbf{a}) = \{\lambda \in \mathbb{C}^n : \mathbf{a} - \lambda \notin B^-\}$$
$$\sigma_b(\mathbf{a}, X) = \sigma_b^+(\mathbf{a}) \cup \sigma_b^-(\mathbf{a})$$

In [8] it is shown that for commuting tuples σ_b is a compact-valued joint spectrum which satisfies a spectral mapping theorem. We note the following inclusions,

$$\sigma_e^+(\mathbf{a}) \subseteq \sigma_b^+(\mathbf{a}) \subseteq \sigma_l(\mathbf{a})$$
$$\sigma_e^-(\mathbf{a}) \subseteq \sigma_b^-(\mathbf{a}) \subseteq \sigma_r(\mathbf{a})$$

Theorem 2.1. *Let $\mathbf{a} = (a_1, \ldots, a_n)$ and $\mathbf{b} = (b_1, \ldots, b_m)$ be tuples of operators on a complex Hilbert space H and let $\mathcal{I}$ be either $B(H)$ or a norm ideal of $B(H)$. Then $\sigma_b((L_\mathbf{a}, R_\mathbf{b}), \mathcal{I}) = S_1 \cup S_2$ where*

$$S_1 = \big(\sigma_l(\mathbf{a}) \times \sigma_b^-(\mathbf{b})\big) \cup \big(\sigma_b^+(\mathbf{a}) \times \sigma_r(\mathbf{b})\big)$$
$$S_2 = \big(\sigma_r(\mathbf{a}) \times \sigma_b^+(\mathbf{b})\big) \cup \big(\sigma_b^-(\mathbf{a}) \times \sigma_l(\mathbf{b})\big)$$

Proof. First we show the inclusion $\sigma_b(L_\mathbf{a}, R_\mathbf{b}) \supseteq S_1 \cup S_2$. Suppose $\lambda \notin \sigma_b(L_\mathbf{a}, R_\mathbf{b})$. For simplicity we assume $\lambda = 0 \in \mathbb{C}^{n+m}$. Then $(L_\mathbf{a}, R_\mathbf{b})$ is Fredholm with finite ascent and finite descent. Using (2.1) and (2.2) we have four possible cases to consider:

(i) $0 \notin (\sigma_l(\mathbf{a}) \times \sigma_r(\mathbf{b}))$ and $0 \notin (\sigma_r(\mathbf{a}) \times \sigma_l(\mathbf{b}))$
(ii) $0 \notin \sigma_l(\mathbf{a}) \times \sigma_r(\mathbf{b})$ and $0 \in (\sigma_r(\mathbf{a}) \backslash \sigma_e^-(\mathbf{a})) \times (\sigma_l(\mathbf{b}) \backslash \sigma_e^+(\mathbf{b}))$
(iii) $0 \notin \sigma_r(\mathbf{a}) \times \sigma_l(\mathbf{b})$ and $0 \in (\sigma_l(\mathbf{a}) \backslash \sigma_e^+(\mathbf{a})) \times (\sigma_r(\mathbf{b}) \backslash \sigma_e^-(\mathbf{b}))$
(iv) $0 \in (\sigma_r(\mathbf{a}) \backslash \sigma_e^-(\mathbf{a})) \times (\sigma_l(\mathbf{b}) \backslash \sigma_e^+(\mathbf{b}))$ and $0 \in (\sigma_l(\mathbf{a}) \backslash \sigma_e^+(\mathbf{a})) \times (\sigma_r(\mathbf{b}) \backslash \sigma_e^-(\mathbf{b}))$

If (i) holds then clearly $0 \notin S_1 \cup S_2$. Suppose (ii) holds. Since $0 \in \sigma_r(\mathbf{a})$ we have $\delta(\mathbf{a}) > 0$. Since $0 \in \sigma_l(\mathbf{b})$ and $\mathbf{b}$ is upper semi-Fredholm we have $\alpha(\mathbf{b}) > 0$. Thus by Theorem 1.4, $\delta(\mathbf{a}) + \alpha(\mathbf{b}) - 1 \leq \delta(L_\mathbf{a}, R_\mathbf{b}) < \infty$. From (ii) we have either $0 \notin \sigma_l(\mathbf{a})$ or $0 \notin \sigma_r(\mathbf{b})$. If $0 \notin \sigma_l(\mathbf{a})$ then $\alpha(\mathbf{a}) = 0$. But this forces $\alpha(\mathbf{a}) = \delta(\mathbf{a}) = 0$ which is a contradiction. Similarly if $0 \notin \sigma_r(\mathbf{b})$ then $\alpha(\mathbf{b}) = \delta(\mathbf{b}) = 0$ which is a contradiction. We conclude that (ii) cannot hold. A similar argument shows that

(iii) cannot hold. If (iv) holds then $\mathbf{a}$ and $\mathbf{b}$ are both Fredholm tuples with non-zero ascent and descent. Since $\mathbf{b}$ is Fredholm we have $\alpha(\mathbf{b}^*) = \delta(\mathbf{b})$ ([8, Proposition 3.5]). Applying Theorem 1.4 we see that $\mathbf{a}$ and $\mathbf{b}$ both have finite ascent and finite descent. Hence $0 \notin S_1 \cup S_2$. This shows that $\sigma_b(L_{\mathbf{a}}, R_{\mathbf{b}}) \supseteq S_1 \cup S_2$.

Next we show the inclusion $\sigma_b(L_{\mathbf{a}}, R_{\mathbf{b}}) \subseteq S_1 \cup S_2$. Suppose $\lambda \notin S_1 \cup S_2$. Again we will assume $\lambda = 0 \in \mathbb{C}^{n+m}$. Using (2.1) and (2.2) we see that $(L_{\mathbf{a}}, R_{\mathbf{b}})$ is Fredholm. To show that $(L_{\mathbf{a}}, R_{\mathbf{b}})$ has finite ascent and descent we consider four possibilities:

(i) $0 \notin (\sigma_l(\mathbf{a}) \times \sigma_r(\mathbf{b}))$ and $0 \notin (\sigma_r(\mathbf{a}) \times \sigma_l(\mathbf{b}))$
(ii) $0 \notin \sigma_l(\mathbf{a}) \times \sigma_r(\mathbf{b})$ and $0 \in (\sigma_r(\mathbf{a})\backslash\sigma_b^-(\mathbf{a})) \times (\sigma_l(\mathbf{b})\backslash\sigma_b^+(\mathbf{b}))$
(iii) $0 \notin \sigma_r(\mathbf{a}) \times \sigma_l(\mathbf{b})$ and $0 \in (\sigma_l(\mathbf{a})\backslash\sigma_b^+(\mathbf{a})) \times (\sigma_r(\mathbf{b})\backslash\sigma_b^-(\mathbf{b}))$
(iv) $0 \in (\sigma_r(\mathbf{a})\backslash\sigma_b^-(\mathbf{a})) \times (\sigma_l(\mathbf{b})\backslash\sigma_b^+(\mathbf{b}))$ and $0 \in (\sigma_l(\mathbf{a})\backslash\sigma_b^+(\mathbf{a})) \times (\sigma_r(\mathbf{b})\backslash\sigma_b^-(\mathbf{b}))$

If (i) holds then we obtain one of the four conditions

$$\alpha(\mathbf{a}) = \delta(\mathbf{a}) = 0 \qquad \alpha(\mathbf{b}^*) = \delta(\mathbf{b}^*) = 0$$
$$\alpha(\mathbf{a}) = \delta(\mathbf{b}^*) = 0 \qquad \alpha(\mathbf{b}^*) = \delta(\mathbf{a}) = 0$$

From Theorem 1.1 and Theorem 1.2, each condition gives $\alpha(L_{\mathbf{a}}, R_{\mathbf{b}}) = \delta(L_{\mathbf{a}}, R_{\mathbf{b}}) = 0$. If (ii) or (iii) holds then we have either $\alpha(\mathbf{a}) = \delta(\mathbf{a}) = 0$ or $\alpha(\mathbf{b}^*) = \delta(\mathbf{b}^*) = 0$. Again this implies $\alpha(L_{\mathbf{a}}, R_{\mathbf{b}}) = \delta(L_{\mathbf{a}}, R_{\mathbf{b}}) = 0$. If (iv) holds then both $\mathbf{a}$ and $\mathbf{b}^*$ are Fredholm tuples with finite non-zero ascent and descent. By Corollary 1.5 we have $\alpha(L_{\mathbf{a}}, R_{\mathbf{b}}) = \delta(L_{\mathbf{a}}, R_{\mathbf{b}}) < \infty$. Hence $0 \notin \sigma_b(L_{\mathbf{a}}, R_{\mathbf{b}})$. This shows $\sigma_b(L_{\mathbf{a}}, R_{\mathbf{b}}) \subseteq S_1 \cup S_2$. □

We use below the standard notation $A \circ B = \{\sum_{i=1}^n \lambda_i\mu_i : \lambda \in A,\ \mu \in B\}$ for subsets A, B of $\mathbb{C}^n$.

Corollary 2.2. *Let $\mathbf{a} = (a_1, \dots, a_n)$ and $\mathbf{b} = (b_1, \dots, b_n)$ be commuting n-tuples of operators on a complex Hilbert space H and let $\mathcal{I}$ be $B(H)$ or a norm ideal of $B(H)$. Let $\mathcal{E} : \mathcal{I} \to \mathcal{I}$ be the elementary operator $\mathcal{E}(u) = a_1ub_1 + \cdots + a_nub_n$. Then*

$$\sigma_b(\mathcal{E}, \mathcal{I}) = \big(\sigma_l(\mathbf{a}) \circ \sigma_b^-(\mathbf{b})\big) \cup \big(\sigma_b^+(\mathbf{a}) \circ \sigma_r(\mathbf{b})\big) \cup \big(\sigma_r(\mathbf{a}) \circ \sigma_b^+(\mathbf{b})\big) \cup \big(\sigma_b^-(\mathbf{a}) \circ \sigma_l(\mathbf{b})\big)$$

Proof. Write $\mathcal{E} = p(L_{\mathbf{a}}, R_{\mathbf{b}})$ where p is the polynomial $p(z_1, \dots, z_n, w_1, \dots, w_n) = \sum_{i=1}^n z_iw_i$. Applying the spectral mapping theorem for σ_b ([8, §4]) we obtain $\sigma_b(\mathcal{E}) = p(\sigma_b((L_{\mathbf{a}}, R_{\mathbf{b}}), \mathcal{I}))$ and so the result follows from Theorem 2.1. □

The Taylor–Browder spectrum of a commuting n-tuple $\mathbf{a} = (a_1, \dots, a_n)$ is

$$\sigma_{Tb}(\mathbf{a}) = \operatorname{acc}(\sigma_T(\mathbf{a})) \cup \sigma_{Te}(\mathbf{a})$$

where σ_T is the Taylor spectrum, σ_{Te} is the essential Taylor spectrum and acc denotes the set of accumulation points. The Taylor–Browder spectrum of $(L_{\mathbf{a}}, R_{\mathbf{b}})$ is readily deduced from [5],

$$\sigma_{Tb}((L_{\mathbf{a}}, R_{\mathbf{b}}), \mathcal{I}) = (\sigma_T(\mathbf{a}) \times \sigma_{Tb}(\mathbf{b})) \cup (\sigma_{Tb}(\mathbf{a}) \times \sigma_T(\mathbf{b}))$$

Application of the spectral mapping theorem for σ_{Tb} ([4]) yields

$$\sigma_b(\mathcal{E}, \mathcal{I}) = (\sigma_T(\mathbf{a}) \circ \sigma_{Tb}(\mathbf{b})) \cup (\sigma_{Tb}(\mathbf{a}) \circ \sigma_T(\mathbf{b}))$$

Combining this with Corollary 2.2 and noting inclusions we obtain the formula

$$\sigma_b(\mathcal{E}, \mathcal{I}) = (\sigma_H(\mathbf{a}) \circ \sigma_b(\mathbf{b})) \cup (\sigma_b(\mathbf{a}) \circ \sigma_H(\mathbf{b}))$$

where $\sigma_H = \sigma_l \cup \sigma_r$ denotes the Harte spectrum.

Acknowledgment

Many thanks to Martin Mathieu for prompting this research and to the Analysis group at TCD for valuable discussions.

References

[1] M. Burgos, A. Kaidi, M. Mbekhta, M. Oudghiri, The descent spectrum and perturbations, *J. Operator Theory* 56(2) (2006), 259–271.

[2] A. Carrillo, C. Hernández, Spectra of constructs of a system of operators, *Proc. Amer. Math. Soc.* 91(3) (1984), 426–432.

[3] R.E. Curto, Spectral theory of elementary operators, in: M. Mathieu (ed.), *Elementary operators and applications* (*Blaubeuren* 1991), *World Sci. Publ., River Edge, NJ* (1992), 3–52.

[4] R.E. Curto, A.T. Dash, Browder spectral systems, *Proc. Amer. Math. Soc.* 103(2) (1988), 407–413.

[5] R.E. Curto, L. Fialkow, The spectral picture of $[L_A, R_B]$, *J. Funct. Anal.* 71 (1987), 371–392.

[6] L. Fialkow, Essential spectra of elementary operators, *Trans. Amer. Math. Soc.* 267(1) (1981), 157–174.

[7] R.E. Harte, Tensor products, multiplication operators and the spectral mapping theorem, *Proc. Royal Irish Acad.* 73A (1973), 285–302.

[8] D. Kitson, Ascent and descent for sets of operators, *Studia Math.* 191(2) (2009), 151–161.

[9] G. Lumer, M. Rosenblum, Linear operator equations, *Proc. Amer. Math. Soc.* 10(1) (1959), 32–41.

[10] M. Mathieu, Elementary operators on prime C^*-algebras, I, *Math. Ann.* 284 (1989), 223–244.

[11] V. Müller, *Spectral Theory of Linear Operators and Spectral Systems in Banach Algebras*, Birkhäuser Verlag, 2007.

Derek Kitson
School of Mathematics
Trinity College
Dublin 2, Ireland
e-mail: dk@maths.tcd.ie

Approximation of Maps on C^*-algebras by Completely Contractive Elementary Operators

Bojan Magajna

Abstract. Let $\overline{\mathrm{E}_1(\mathrm{A})}^{p.n.}$ be the closure in the point-norm topology of the set of all completely contractive elementary operators on a C^*-algebra A. If $\psi \leq \phi$ are completely positive maps on A and $\phi \in \overline{\mathrm{E}_1(\mathrm{A})}^{p.n.}$, then $\psi \in \overline{\mathrm{E}_1(\mathrm{A})}^{p.n.}$. A completely positive contraction ϕ on a von Neumann algebra R is in $\overline{\mathrm{E}_1(\mathrm{R})}^{p.n.}$ if and only if the normal and the singular part of ϕ are both in $\overline{\mathrm{E}_1(\mathrm{R})}^{p.n.}$. Maps on R admitting pointwise approximation by *sequences* of elementary complete contractions may have additional properties that are not shared by all maps in $\overline{\mathrm{E}_1(\mathrm{R})}^{p.n.}$. A specific example on $\mathrm{B}(\mathcal{H})$ is also studied.

Mathematics Subject Classification (2000). Primary 46L07; Secondary 47B47.

Keywords. Elementary operators, complete contractions, C^*-algebras, von Neumann algebras.

1. Introduction and notation

Given a Banach algebra A, we may ask which operators on A can be approximated by elementary operators, either uniformly or pointwise. Clearly all such operators must preserve ideals of A (where by an ideal we mean a closed two-sided ideal) and for C^*-algebras this necessary condition is also sufficient for the pointwise approximation [2], [13]. Here by an operator we shall always mean a linear bounded operator.

A variant of the above problem is to characterize all Banach or all C^*-algebras A such that every operator on A that preserves ideals can be approximated uniformly by elementary operators. Separable C^*-algebras with this property are characterized in [17] (they are just finite direct sums of homogeneous C^* algebras

Partially supported by the Ministry of Science and Education of Slovenia.

The author is grateful to Aleksej Turnšek and Milan Hladnik for their remarks concerning the paper.

R.E. Curto and M. Mathieu (eds.), *Elementary Operators and Their Applications*, Operator Theory: Advances and Applications 212, DOI 10.1007/978-3-0348-0037-2_3, © Springer Basel AG 2011

of finite type), but for nonseparable C^*-algebras and more general Banach algebras the problem has not yet been studied, as far as we know. Since the set of all elementary operators on a (unital) C^*-algebra A coincides with the range of the natural map from the algebraic tensor product $A \otimes_Z A$ (over the center Z of A) to A, this question is in some sense dual to the problem of characterizing all C^* algebras A for which the multiplication map is (completely) isometric when $A \otimes_Z A$ is equipped with the Haagerup (semi-)norm. The latter problem received much attention (see [8], [20], [6], [1], [14], [2] and the references there) and a definitive solution has been obtained recently by Somerset [21] and Archbold, Somerset and Timoney [3].

On a C^*-algebra every elementary operator ψ is a *completely bounded* map in the sense that $\|\psi\|_{\text{cb}} = \sup_n \|\psi_n\| < \infty$, where for each $n \in \mathbb{N}$ we denote by ψ_n the map on $\mathrm{M}_n(A)$ obtained by applying ψ entrywise. So, in this case we may ask a sharper question: which complete contractions ϕ on A (that is, $\|\phi\|_{\text{cb}} \leq 1$) that preserve closed two-sided ideals, can be approximated pointwise by elementary complete contractions (that is, by completely contractive elementary operators)? We note that the approximation on singletons is always possible [15], the problem is the simultaneous approximation on more general finite subsets of a C^*-algebra. It turns out that this question is also related to C^*-tensor products of C^*-algebras [16]. (However, here we will not study the relation to tensor products of C^*-algebras and no knowledge of the theory of tensor products of C^*-algebras is needed to read this paper.)

Since the Calkin algebra $\mathrm{C}(\mathcal{H}) = \mathrm{B}(\mathcal{H})/\mathrm{K}(\mathcal{H})$ of a separable Hilbert space is algebraically simple, it is a consequence of the classical Jacobson density theorem that every linear operator ϕ on $\mathrm{C}(\mathcal{H})$ coincides on each finite-dimensional subspace V of $\mathrm{C}(\mathcal{H})$ with some elementary operator ψ_V. But if $\|\phi\|_{\text{cb}} < 1$, in general ψ_V can not be chosen to be completely contractive; namely, as observed in [16], there is a lifting obstruction for such an approximation. Thus, the requirement that the approximands should be completely contractive is essential for the character of the approximation problem studied here.

In Section 2 we will reformulate the above question and provide a different approach to a part of the main result of [16].

In Section 3 we will deduce some consequences for completely positive maps of the result of Section 2, needed later. In particular, for completely positive contractions $\psi \leq \phi$ on a C^*-algebra A we show that $\phi \in \overline{\mathrm{E}_1(\mathrm{A})}^{p.n.}$ implies $\psi \in \overline{\mathrm{E}_1(\mathrm{A})}^{p.n.}$, where $\overline{\mathrm{E}_1(\mathrm{A})}^{p.n.}$ denotes the point-norm closure of the set $\mathrm{E}_1(A)$ of all elementary operators on a C^*-algebra A that can be represented by tensors in $A \otimes A$ with the Haagerup norm at most 1. (By Theorem 2.1 below this is the same as the point-norm closure of all completely contractive elementary operators on A.)

Using these we will show in Section 4 that a completely positive contraction on a von Neumann algebra R is in $\overline{\mathrm{E}_1(\mathrm{R})}^{p.n.}$ if and only if its normal and its singular part are both in $\overline{\mathrm{E}_1(\mathrm{R})}^{p.n.}$. We will also note that if a map ϕ on R can be approximated pointwise by a *sequence* of completely contractive elementary operators, then its second adjoint ϕ^{**} on R^{**} is a module homomorphism over the center of R^{**}, and that this property does not hold in general for all maps in $\overline{\mathrm{E}_1(\mathrm{R})}^{p.n.}$.

Recall that every bounded linear map ϕ on $\mathrm{B}(\mathcal{H})$ can be decomposed uniquely as $\phi = \phi_{\text{sing}} + \phi_{\text{nor}}$, where ϕ_{nor} is normal (meaning weak* continuous) and ϕ_{sing} is *singular* (which is equivalent to $\phi_{\text{sing}}(\mathrm{K}(\mathcal{H})) = 0$). If ϕ is completely contractive, Voiculescu's theorem [4], [23] can be used to approximate the singular part ϕ_{sing} by two-sided multiplications of the form $x \mapsto axb$, where a and b are contractions on $\mathcal{H}$, and the approximation problem can be reduced to normal maps. It is well known that all normal completely bounded maps on $\mathrm{B}(\mathcal{H})$ ($\mathcal{H}$ separable) are of the form $\phi(x) = \sum_{i=1}^{\infty} a_i^* x b_i$, where a_i and b_i are elements of $\mathrm{B}(\mathcal{H})$ such that the two infinite columns $a = (a_1, a_2, \ldots)^T$ and $b = (b_1, b_2, \ldots)^T$ represent bounded operators from $\mathcal{H}$ into $\mathcal{H}^\infty$ (in other words, the two sums $\sum a_i^* a_i$ and $\sum b_i^* b_i$ are convergent in the strong operator topology). Occasionally we shall denote such a map ϕ by $a^* \odot b$. But such maps do not always preserve compact operators: for example, if e_{ij} are the matrix units of $\mathrm{B}(\mathcal{H})$ (with respect to some orthonormal basis of $\mathcal{H}$), the map $x \mapsto \sum e_{i1} x e_{1i}$ sends the rank 1 projection e_{11} into the identity operator, which is not compact. In Section 5 we will consider a special class of normal complete contractions on $\mathrm{B}(\mathcal{H})$ that do preserve the ideal $\mathrm{K}(\mathcal{H})$ of compact operators and show that they can not always be approximated pointwise by *sequences* of elementary complete contractions. Since $\mathrm{B}(\mathcal{H})$ is not separable in norm, this does not imply that such maps are not in $\overline{\mathrm{E}_1(\mathrm{B}(\mathcal{H}))}^{p.n.}$ and the question if all complete contractions on $\mathrm{B}(\mathcal{H})$ that preserve $\mathrm{K}(\mathcal{H})$ are in $\overline{\mathrm{E}_1(\mathrm{B}(\mathcal{H}))}^{p.n.}$ is not answered completely in this paper. Of course, the approximation in the point-weak* topology is always possible in $\mathrm{B}(\mathcal{H})$. More generally it follows from results of Chatterjee and Smith [6] that each complete contraction on an injective von Neumann algebra R which preserves weak* closed two-sided ideals of R can be approximated by elementary complete contractions in the point-weak* topology.

We shall denote by Φ the universal representation of a von Neumann (or a C^*-algebra) R. Then the weak* closure $\overline{\Phi(R)}$ of $\Phi(R)$ (in $\mathrm{B}(\mathcal{K})$, where $\mathcal{K}$ is the Hilbert space of Φ) is naturally identified with the bidual R^{**} of R. Each functional in R^* can be extended uniquely to a normal functional on R^{**} and the map $\Phi^{-1} : \Phi(R) \to R$ has the weak* continuous extension $\overline{\Phi^{-1}} : R^{**} \to R$ (or to $\overline{R}$ if R is merely a C^*-algebra represented on a Hilbert space). Since $\overline{\Phi^{-1}}$ is a $*$-homomorphism, the kernel of $\overline{\Phi^{-1}}$ is a weak* closed ideal in R^{**}, hence of the form $P^\perp R^{**}$ for a central projection $P \in R^{**}$. Since the map ϕ^{**} on R^{**} is weak* continuous, the composite $\overline{\phi} := \overline{\Phi^{-1}}\phi^{**}$ is also weak* continuous and is the unique weak* continuous extension of $\phi\Phi^{-1}$ to a map $R^{**} \to R$. The normal and the singular part of ϕ are then given by $\phi_{\text{nor}}(x) = \overline{\phi}(P\Phi(x))$ and $\phi_{\text{sing}}(x) = \overline{\phi}(P^\perp\Phi(x))$ $(x \in R)$. More explanation on this can be found, e.g., in [12, Section 10.1].

2. A preliminary result

The following result is a reformulation of a part of [16, Theorem 2.1] (with the additional observation, that it suffices to consider the universal representations). We will give a proof, different from that in [16].

Theorem 2.1. *Let ϕ be a complete contraction on a C^*-algebra A such that $\phi(J) \subseteq J$ for each closed two-sided ideal J of A. Then ϕ can be approximated pointwise by a net of elementary complete contractions ψ_k if and only if for each representation $\pi : A \to \mathrm{B}(\mathcal{H})$ and all finite subsets $\{x_1, \ldots, x_n\}$ of A and $\{x'_1, \ldots, x'_n\}$ of $\pi(A)'$ (the commutant of $\pi(A)$) the inequality*

$$\Big\|\sum_{i=1}^{n} x'_i \pi(\phi(x_i))\Big\| \leq \Big\|\sum_{i=1}^{n} x'_i \pi(x_i)\Big\| \tag{2.1}$$

holds. Moreover, it suffices that (2.1) holds for the universal representation of A and the approximating operators ψ_k can be chosen to be of the form $\psi_k(x) = \sum_i a_{i,k} x b_{i,k}$ $(a_{i,k}, b_{i,k} \in A)$ with $\|\sum_i a_{i,k} a^*_{i,k}\| \leq 1$, $\|\sum_i b^*_{i,k} b_{i,k}\| \leq 1$ (that is, the Haagerup norms of tensors corresponding to the maps ψ_k are at most 1).

Proof. Suppose first that ϕ is an elementary operator on A with $\|\phi\|_{\mathrm{cb}} < 1$. Let $\pi : A \to \mathrm{B}(\mathcal{H})$ be a representation of A and ϕ_π the elementary operator induced by ϕ on the weak* closure R of $\pi(A)$. Since $\|\phi_\pi\|_{\mathrm{cb}}$ is equal to the norm of the tensor corresponding to ϕ_π in the central Haagerup tensor product $R \otimes^h_Z R$ (see [6]), we have that ϕ_π is of the form $\phi_\pi(x) = \sum_j c^*_j x d_j$ $(x \in R)$, where c_j, $d_j \in R$ and the columns $c := (c_1, \ldots, c_m)^T$ and $d := (d_1, \ldots, d_m)^T$ have norms less than 1. Then for any $x_i \in A$ and $x'_i \in \pi(A)'$ $(i = 1, \ldots, n)$ we compute (denoting by $x^{(m)}$ the $m \times m$ diagonal matrix with x along the diagonal) that

$$\begin{aligned}\|\textstyle\sum_i x'_i \pi(\phi(x_i))\| = \|\sum_i x'_i \phi_\pi(\pi(x_i))\| &= \|\textstyle\sum_i x'_i \sum_j c^*_j \pi(x_i) d_j\| \\ &= \|c^*(\textstyle\sum_i x'_i \pi(x_i))^{(m)} d\| \leq \|c\| \|\sum_i x'_i \pi(x_i)\| \|d\| \\ &\leq \|\textstyle\sum_i x'_i \pi(x_i)\|.\end{aligned}$$

This proves (2.1) in the case ϕ is elementary with $\|\phi\|_{\mathrm{cb}} < 1$, the general case follows by an approximation.

Suppose now that (2.1) holds for the universal representation of A. We will regard A as a C^*-subalgebra of its universal von Neumann envelope $A'' = A^{**}$ (and so regard the universal representation as the inclusion). So the condition (2.1) simply says that

$$\Big\|\sum_{i=1}^{n} \phi(x_i) x'_i\Big\| \leq \Big\|\sum_{i=1}^{n} x_i x'_i\Big\|$$

for all finite subsets $\{x_1, \ldots, x_n\}$ of A and $\{x'_1, \ldots, x'_n\}$ of A', which means that ϕ extends to a contractive A'-bimodule homomorphism $\tilde{\phi}$ of the C^*-algebra $\overline{\overline{AA'}}$ generated by A and A' inside $\mathrm{B}(\mathcal{K})$, where $\mathcal{K}$ is the Hilbert space of the universal representation of A. Since every normal state on A'' (that is, every state on A) is a vector state [12, Section 10.1], every finite subset of $\mathcal{K}$ is contained in a cyclic subspace for A' by [20, Lemma 2.3] and it follows that $\tilde{\phi}$ is completely contractive by [20, Theorem 2.1 and Remark 2.2]. Then by an appropriate variant of the Wittstock extension theorem [5], [18, p. 116] ϕ can be extended to a completely contractive A'-bimodule homomorphism ψ on $\mathrm{B}(\mathcal{K})$. By [8] the space $CB_{A'}(\mathrm{B}(\mathcal{K}))_{A'}$ of all completely bounded A'-bimodule homomorphisms on $\mathrm{B}(\mathcal{K})$ can be identified with the second dual $(A \overset{h}{\otimes} A)^{**}$ of the Haagerup tensor product $A \overset{h}{\otimes} A$ completely isometrically and weak* homeomorphically by the map which extends the well-known complete isometry $A \overset{h}{\otimes} A \to \mathrm{B}(\mathcal{K})$. (Perhaps a simpler proof of this is in [7]: on the level of preduals the corresponding map is the result of the following identifications: $\mathcal{K}^* \overset{h}{\otimes}_{A'} \mathrm{B}(\mathcal{K}) \overset{h}{\otimes}_{A'} \mathcal{K} = \mathcal{K}^* \overset{eh}{\otimes}_{A'} (\mathcal{K} \overset{eh}{\otimes} \mathcal{K}^*) \overset{eh}{\otimes}_{A'} \mathcal{K} = (\mathcal{K}^* \overset{h}{\otimes}_{A'} \mathcal{K}) \overset{eh}{\otimes} (\mathcal{K}^* \overset{h}{\otimes}_{A'} \mathcal{K}) = A^* \overset{eh}{\otimes} A^*$. Here $\overset{eh}{\otimes}$ denotes the extended Haagerup tensor product and we have used some relations from [7].) It follows that ψ is the point-weak* limit of a net ψ_k of elementary complete contractions on $\mathrm{B}(\mathcal{K})$ of the form $\psi_k(x) = \sum_i a_{i,k} x b_{i,k}$, where $a_{i,k}, b_{i,k} \in A$, for each k the sum is finite and $\sum_i a_{i,k} a_{i,k}^* \leq 1$, $\sum_i b_{i,k}^* b_{i,k} \leq 1$. Since such elementary operators map A into A and each functional in A^* extends to a normal functional on $\mathrm{B}(\mathcal{K})$, it follows that the restrictions $\psi_k|A$ converge to $\phi = \psi|A$ in the point-weak topology. Since the point-weak and the point-norm topology have the same continuous linear functionals, it follows by a well-known convexity argument that ϕ can be approximated by elementary complete contractions of the required form. □

3. Pointwise approximation of completely positive maps by completely contractive elementary operators

Proposition 3.1. *Let ϕ and ψ be completely positive maps on a C^*-algebra A such that $\psi \leq \phi$ (that is, $\phi - \psi$ is completely positive). If $\phi \in \overline{\mathrm{E}_1(\mathrm{A})}^{p.n.}$, then $\psi \in \overline{\mathrm{E}_1(\mathrm{A})}^{p.n.}$.*

Proof. We shall use Theorem 2.1. Given a representation π of A and finite subsets $\{x_1, \ldots, x_n\}$ of A and $\{x'_1, \ldots, x'_n\}$ of $\pi(A)'$, we denote by x the $n \times n$ matrix which has the first row $(x_1, \ldots, x_n)$ and the remaining rows all 0, and by x' the matrix with the first column $(x'_1, \ldots, x'_n)^T$ and the remaining columns 0. We denote by π, ϕ and ψ also the amplifications of these maps to the matrix algebra $\mathrm{M}_n(A)$. Then, using the Schwarz inequality for completely positive maps, we compute that

$$\begin{aligned}
\|\textstyle\sum_j \pi(\psi(x_j))x'_j\|^2 &= \|\textstyle\sum_{i,j} {x'_i}^* \pi(\psi(x_i)^*\psi(x_j))x'_j\| = \|{x'}^*\pi(\psi(x)^*\psi(x))x'\| \\
&\leq \|{x'}^*\pi(\psi(x^*x))x'\| \qquad\qquad \leq \|{x'}^*\pi(\phi(x^*x))x'\| \\
&= \|\textstyle\sum_{i,j} \pi(\phi(x_i^*x_j)){x'}_i^*x'_j\| \qquad \leq \|\textstyle\sum_{i,j} \pi(x_i^*x_j){x'}_i^*x'_j\| \\
&= \|\textstyle\sum_j \pi(x_j)x'_j\|^2,
\end{aligned}$$

where for the last inequality we have used the condition (2.1) for the map ϕ and the subsets $\{x_i^*x_j : i,j = 1,\ldots,n\}$ and $\{{x'}_i^*x'_j : i,j = 1,\ldots,n\}$. □

Given a map ψ on a C^*-algebra A and elements $a, b \in A$, we denote by $a\psi b$ the map on A defined by $(a\psi b)(x) = a\psi(x)b$.

Lemma 3.2. *Let ψ and θ be completely positive maps on a unital C^*-algebra A and let $h, k \in A^+$ be such that $h^2 + k^2 \leq 1$. If for all $t \in (0,1)$ the two maps $h^t\psi h^t$ and $k^t\theta k^t$ are in $\overline{\mathrm{E}_1(\mathrm{A})}^{p.n.}$ then $\phi := h\psi h + k\theta k$ is also in $\overline{\mathrm{E}_1(\mathrm{A})}^{p.n.}$.*

Proof. Let $t \in (0,1)$. For each $x \in A$ we may write

$$\phi(x) = \begin{bmatrix} h^{1-t} & k^{1-t} \end{bmatrix} \begin{bmatrix} h^t\psi(x)h^t & 0 \\ 0 & k^t\theta(x)k^t \end{bmatrix} \begin{bmatrix} h^{1-t} \\ k^{1-t} \end{bmatrix}.$$

If ψ_k and θ_k are nets of maps in $\mathrm{E}_1(A)$ converging pointwise to $h^t\psi h^t$ and $k^t\theta k^t$, respectively, then the elementary operators $\phi_k := h^{1-t}\psi_k h^{1-t} + k^{1-t}\theta_k k^{1-t}$ converge pointwise to ϕ. Further, $\|\phi_k\|_{\mathrm{cb}} \leq \|h^{2(1-t)} + k^{2(1-t)}\|$, so now it suffices to observe that $\limsup_{t\to 0} \|h^{2(1-t)} + k^{2(1-t)}\| \leq 1$. This follows from the relations $\lim_{t\to 0} \|h^{2(1-t)} - h^2\| = 0 = \lim_{t\to 0} \|k^{2(1-t)} - k^2\|$, which can be proved by the functional calculus. □

Lemma 3.3. *Let ψ be a completely positive contraction on a unital C^*-algebra A and $h \in A$, $0 \leq h \leq 1$. Suppose that $h\psi h \in \overline{\mathrm{E}_1(\mathrm{A})}^{p.n.}$. Then $h^t\psi h^t \in \overline{\mathrm{E}_1(\mathrm{A})}^{p.n.}$ for all $t > 0$; more generally, $f(h)^*\psi f(h) \in \overline{\mathrm{E}_1(\mathrm{A})}^{p.n.}$ for any continuous function f on the spectrum of h such that $\|f\| \leq 1$ and (if h is not invertible) $f(0) = 0$.*

Proof. Since $h\psi h \in \overline{\mathrm{E}_1(\mathrm{A})}^{p.n.}$, $h\psi h$ satisfies the condition (2.1) of Theorem 2.1, hence for each positive constant $c \in \mathbb{R}$ the map $\theta_c := (h+c)^{-1}hf(h)^*\psi f(h)h(h+c)^{-1}$ satisfies the following condition: for every unital representation π of A and every finite subsets $\{x_1, \ldots, x_n\}$ of A and $\{x'_1, \ldots, x'_n\}$ of $\pi(A)'$ the inequality

$$\left\|\sum_i \pi(\theta_c(x_i))x'_i\right\| \leq \kappa_c \left\|\sum_i \pi(x_i)x'_i\right\|$$

holds, where $\kappa_c = \|f(h)h(h+c)^{-1}\|^2$. This means that the map $\sum_i \pi(x_i)x'_i \mapsto \sum_i \pi(\theta_c(x_i))x'_i$ extends to a bounded map $\tilde{\theta}_c$ on the C^*-algebra $\overline{\pi(A)\pi(A)'}$. Since ψ is completely positive, it is not hard to verify that

$$\tilde{\theta}_c\left(\left(\sum_i \pi(x_i)x'_i\right)^*\left(\sum_i \pi(x_i)x'_i\right)\right) \geq 0,$$

hence $\tilde{\theta}_c$ is positive. A similar argument shows that $\tilde{\theta}_c$ is completely positive, so $\|\tilde{\theta}_c\|_{\mathrm{cb}} = \|\tilde{\theta}_c(1)\| \leq \|\theta_c(1)\| = \|f(h)h(h+c)^{-1}\|^2 \leq 1$. Therefore

$$\left\|\sum_i \pi(\theta_c(x_i))x'_i\right\| \leq \left\|\sum_i \pi(x_i)x'_i\right\|$$

for any finite sets (x_i) and (x'_i) and representation π as in Theorem 2.1, hence $\theta_c \in \overline{\mathrm{E}_1(\mathrm{A})}^{p.n.}$. This holds for every choice of c, so it suffices now to observe that θ_c tends uniformly to $f(h)^*\psi f(h)$ as $c \to 0$ since $\lim_{c\to 0} \|f(h)h(h+c)^{-1} - f(h)\| = 0$ by the functional calculus. □

With a similar proof as that of Lemma 3.3 we can show the following proposition.

Proposition 3.4. *Let A be a unital C^*-algebra. If a completely positive map ϕ is in $\overline{\mathrm{E}_1(\mathrm{A})}^{p.n.}$, then ϕ is in the pointwise closure of a net of elementary completely positive contractions σ_k. If, in addition, ϕ is unital, σ_k can be chosen to be unital.*

Proof. Let $h = \sqrt{\phi(1)}$. If h is invertible, we consider the unital completely positive map $\psi = h^{-1}\phi h^{-1}$ and show that it satisfies (2.1) by using complete positivity (in the same way as in the proof of Lemma 3.3). So by Theorem 2.1 there is a net $\psi_k = a_k^* \odot b_k$ in $\mathrm{E}_1(A)$ converging pointwise to ψ such that $\|a_k\|, \|b_k\| \leq 1$. Since

$$\|a_k - b_k\|^2 = \|(a_k - b_k)^*(a_k - b_k)\| \leq 2\|1 - \mathrm{Re}(a^*b)\| \to 0, \qquad (3.1)$$

the net of completely positive contractions $\theta_k = a_k^* \odot a_k$ also converges pointwise to ψ. Hence the net of maps $\sigma_k := h\theta_k h = c_k^* \odot c_k$, where $c_k := a_k h$, converges to $h\psi h = \phi$. If $\phi(1) = 1$,we may replace each σ_k by the unital map $\tau_k(x) = \sigma_k(x) + \sqrt{1 - c_k^* c_k}\, x \sqrt{1 - c_k^* c_k}$. In general (if h is not invertible), we consider first the maps $(1-t)\phi + t\,\mathrm{id}$ (where id is the identity map on A) and then let $t \to 0$. □

4. Stability of maps in $\overline{\mathrm{E}_1(\mathrm{R})}^{p.n.}$ under the decomposition into the singular and the normal part

The following lemma is perhaps well known.

Lemma 4.1. *Let $\phi : R \to S$ be a completely positive map between von Neumann algebras and denote $\phi(1) = h^2$, where $h \geq 0$. Then there exists a unique completely positive map $\psi : R \to S$ such that $\psi(1)$ is the range projection p of h and $\phi(x) = h\psi(x)h$ for all $x \in R$. Moreover, ψ is normal (singular) if and only if ϕ is normal (singular).*

Proof. We may assume that $\|\phi\|_{\mathrm{cb}} = 1$, so that $\|h\| = 1$. The sequence of completely positive maps

$$\psi_n(x) = \left(h + \frac{1}{n}\right)^{-1} \phi(x) \left(h + \frac{1}{n}\right)^{-1} \quad (n \in \mathbb{N})$$

is bounded since $\|\psi_n\|_{\mathrm{cb}} = \|\psi_n(1)\| \leq 1$, hence it has a limit point ψ in the point-weak* topology. From $\phi(x) = (h + \frac{1}{n})\psi_n(x)(h + \frac{1}{n})$ we conclude that $\phi(x) = h\psi(x)h$ for all $x \in R$. Since the sequence $\psi_n(1) = (h + \frac{1}{n})^{-1}h^2(h + \frac{1}{n})^{-1}$ converges strongly to the range projection p of h, we have that $\psi(1) = p$. This proves the existence of ψ. The uniqueness of ψ is a consequence of the fact that the range of ψ must be contained in pSp (since $0 \leq \psi(x) \leq \psi(1) = p$ if $0 \leq x \leq 1$). Namely, if $\tilde{\psi}$ is another such map, then the difference $\theta := \tilde{\psi} - \psi$ satisfies $h\theta(x)h = \phi(x) - \phi(x) = 0$. But, since $\theta(x) = p\theta(x)p$ and the operator $h = ph = hp$ is injective on the range of p, this implies that $\theta(x) = 0$ for all (positive) $x \in R$.

We shall now use the notation from the last paragraph of the Introduction. From $\phi = h\psi h$ we have that $\overline{\phi} = h\overline{\psi}h$, hence, if ϕ is normal, then $h\psi(x)h = \phi(x) =$

$\overline{\phi}(P\Phi(x)) = h\overline{\psi}(P\Phi(x))h$ for all $x \in R$. Since ψ and $\overline{\psi}$ both have their ranges contained in pSp, this implies that $\psi(x) = \overline{\psi}(\Phi(x)P)$ $(x \in R)$, so ψ is normal. The converse and the statement about singular maps is proved similarly. □

Corollary 4.2. *A completely positive contraction ϕ on a von Neumann algebra R is in $\overline{\mathrm{E}_1(\mathrm{R})}^{p.n.}$ if and only if its normal part ϕ_{nor} and its singular part ϕ_{sing} are both in $\overline{\mathrm{E}_1(\mathrm{R})}^{p.n.}$.*

Proof. If $\phi \in \overline{\mathrm{E}_1(\mathrm{R})}^{p.n.}$, Proposition 3.1 implies that ϕ_{nor}, $\phi_{\mathrm{sing}} \in \overline{\mathrm{E}_1(\mathrm{R})}^{p.n.}$. To prove the converse, by Lemma 4.1 we write $\phi_{\mathrm{nor}} = h\psi h$ and $\phi_{\mathrm{sing}} = k\theta k$ where ψ is normal, θ is singular, both are completely positive contractions on R and $h, k \in R^+$ satisfy $h^2 = \phi_{\mathrm{nor}}(1)$, $k^2 = \phi_{\mathrm{sing}}(1)$, hence $h^2 + k^2 = \phi(1) \leq 1$. Since ϕ_{nor} and ϕ_{sing} are in $\overline{\mathrm{E}_1(\mathrm{R})}^{p.n.}$, it follows now from Lemma 3.3 that $h^t\psi h^t$ and $k^t\theta k^t$ are in $\overline{\mathrm{E}_1(\mathrm{R})}^{p.n.}$ for all $t > 0$. Then Lemma 3.2 implies that $\phi \in \overline{\mathrm{E}_1(\mathrm{R})}^{p.n.}$. □

The author does not know if Corollary 4.2 can be extended to all completely contractive maps. Note, however, that if ϕ is a point-norm limit of a *sequence* of elementary operators (or more generally, of a sequence of normal maps) ϕ_k on a von Neumann algebra R, then ϕ is necessarily normal. Indeed, for each ω in the predual R_* of R the sequence $(\omega \circ \phi_k)$ in R_* converges weakly to $\omega \circ \phi$, but R_* is weakly sequentially complete [22, 5.2], so $\omega \circ \phi \in R_*$ and ϕ is normal. Here is a somewhat related result.

Proposition 4.3. *If ϕ is a bounded linear map on a von Neumann algebra R which preserves norm closed two-sided ideals of R, then ϕ_{nor} and ϕ_{sing} also preserve such ideals.*

Proof. We shall use again the notation from the last paragraph of the Introduction. If J is a norm closed ideal in R, then J^{**} coincides with the weak* closure of $\Phi(J)$ in R^{**}, hence is of the form $J^{**} = pR^{**}$ for a central projection $p \in R^{**}$. From $\phi(J) \subseteq J$ we have that $\phi^{**}(J^{**}) \subseteq J^{**}$, hence $\phi^{**}(py) = p\phi^{**}(py)$ for all $y \in R^{**}$. Let $x \in J$ and apply the last equality to $y = P\Phi(x) = Pp\Phi(x) = pP\Phi(x)$ to conclude that $\phi_{\mathrm{nor}}(x) = \overline{\phi}(P\Phi(x)) = \overline{\Phi^{-1}}(\phi^{**}(pP\Phi(x))) = \overline{\Phi^{-1}}(p\phi^{**}(pP\Phi(x))) \in \overline{\Phi^{-1}}(J^{**}) \subseteq J$, since $\overline{\Phi^{-1}}$ is weak* continuous and $\Phi^{-1}(\Phi(J)) = J$. Then $\phi_{\mathrm{sing}}(J) \subseteq J$ since $\phi_{\mathrm{sing}} = \phi - \phi_{\mathrm{nor}}$. □

Proposition 4.4. *If a bounded linear map ϕ on a von Neumann algebra R is a point-norm limit of a sequence of elementary operators on R then ϕ^{**} is a module homomorphism over the center of R^{**}. This conclusion does not hold necessarily if we assume only that $\phi \in \overline{\mathrm{E}_1(\mathrm{R})}^{p.n.}$.*

Proof. If ϕ_k is a sequence of elementary operators converging pointwise to ϕ, then for each functional $\rho \in R^*$ the sequence $(\phi_k^*(\rho))$ converges to $\phi^*(\rho) \in R^*$ in the weak* topology, hence, since R is a Grothendieck space [19], the convergence is in the weak topology of R^*. This means that the maps ϕ_k^{**} on R^{**} converge to ϕ^{**} in the point-weak* topology. Note that each ϕ_k^{**} is an elementary operator on

R^{**}, hence a module map over the center Z of R^{**}, and that $\rho z \in R^*$ for each $z \in R^{**}$ and $\rho \in R^*$. (Here ρz is defined by $(\rho z)(x) = \overline{\rho}(zx)$, where $\overline{\rho}$ is the unique weak* continuous extension of ρ to a normal functional on R^{**}, and then ρz can be regarded as a normal functional on R^{**}.) For each $y \in R^{**}$ and $z \in Z$ we compute that

$$\langle \phi^{**}(zy), \rho\rangle = \lim_{k\to\infty} \langle \phi_k^{**}(zy), \rho\rangle = \lim_{k\to\infty} \langle z\phi_k^{**}(y), \rho\rangle = \langle \phi^{**}(y), \rho z\rangle = \langle z\phi^{**}(y), \rho\rangle.$$

Thus, ϕ^{**} is a Z-module map.

To show that the approximation by general nets is not sufficient for the above conclusion, let ω_{ξ_k} be a net of vector states on $\mathrm{B}(\mathcal{H})$ (ξ_k a unit vector in $\mathcal{H}$) weak* converging to a non-normal state ω (by a result of Glimm [10] the weak* closure of vector states on $\mathrm{B}(\mathcal{H})$ consists precisely of states of the form $t\rho + (1-t)\theta$, where $t \in [0,1]$, ρ is a vector state and θ is a state annihilating $\mathrm{K}(\mathcal{H})$). Let $x_0 \in \mathrm{K}(\mathcal{H})$ be the rank 1 operator $\eta \otimes \eta^*$, where $\eta \in \mathcal{H}$ is a unit vector. Then the net (ϕ_k) of complete contractions on $\mathrm{B}(\mathcal{H})$ defined by $\phi_k(x) := \omega_{\xi_k}(x)x_0$ ($x \in \mathrm{B}(\mathcal{H})$) converges pointwise to the map ϕ defined by $\phi(x) = \omega(x)x_0$. Note that each ϕ_k is an elementary operator (which acts as $x \mapsto (\eta \otimes \xi_k^*)x(\xi_k \otimes \eta^*)$). Nevertheless, ϕ^{**} is not a module map over the center Z of $\mathrm{B}(\mathcal{H})^{**}$. To see this, we shall again use the notation from the last paragraph of the Introduction, with $R = \mathrm{B}(\mathcal{H})$. If ϕ^{**} were a Z-module map, then in particular $\phi^{**}(Py) = P\phi^{**}(y)$ for all $y \in \mathrm{B}(\mathcal{H})^{**}$, hence the map $\overline{\phi} = \overline{\Phi^{-1}}\phi^{**}$ would satisfy $\overline{\phi}(Py) = \overline{\Phi^{-1}}(P\phi^{**}(y)) = \overline{\Phi^{-1}}(\phi^{**}(y) = \overline{\phi}(y)$. But then ϕ would be weak* continuous on $\mathrm{B}(\mathcal{H})$ [12, p. 721], a contradiction. □

5. An example on $\mathrm{B}(\mathcal{H})$

In this section let $\mathcal{H}$ be a separable Hilbert space and $\mathrm{K}(\mathcal{H})$ the ideal of compact operators in $\mathrm{B}(\mathcal{H})$.

As noted already in the Introduction, a general normal complete contraction on $\mathrm{B}(\mathcal{H})$ does not necessarily preserve the ideal $\mathrm{K}(\mathcal{H})$ and therefore can not be approximated pointwise in norm by elementary operators. However, for every orthogonal sequence of projections $p_i \in \mathrm{B}(\mathcal{H})$ the completely positive contraction ϕ defined by

$$\phi(x) = \sum_{i=1}^{\infty} p_i x p_i \quad (x \in \mathrm{B}(\mathcal{H})) \tag{5.1}$$

preserves $\mathrm{K}(\mathcal{H})$ for ϕ is just the block-diagonal compression, that is, it sends x to the direct sum of operators $p_i x p_i$. Although the finite partial sums $\sum_{i=1}^n p_i x p_i$ usually do not converge in norm to the infinite sum in (5.1), this does not imply that ϕ can not be approximated pointwise in norm by elementary complete contractions. In fact, if the ranks of the projections p_i are bounded, the range of ϕ is contained in a nuclear C^* algebra and $\phi \in \overline{\mathrm{E}_1(\mathrm{B}(\mathcal{H}))}^{p.n.}$ by [16]. On the other hand, if the ranks of the projections p_i (are finite, but) increase to infinity (or, if infinitely many of the p_i's have infinite rank), then the nuclearity argument does not apply. For simplicity we assume that $\sum p_i = 1$, so that ϕ is unital. The range of ϕ is a von Neumann algebra, which we denote by B_ϕ, and ϕ is an idempotent B_ϕ-bimodule map. We suspect that such maps are critical for the approximation problem on $\mathrm{B}(\mathcal{H})$ in the sense that if all maps of the form (5.1) can be approximated pointwise by completely contractive elementary operators then the same holds for all completely positive contractions that preserve $\mathrm{K}(\mathcal{H})$. We show first that, although ϕ is a B_ϕ-bimodule map, ϕ in general can not be approximated pointwise by completely contractive elementary operators which are B_ϕ-module maps. For this, we need the following finite-dimensional lemma.

Lemma 5.1. *Let $M := \mathrm{M}_n(\mathbb{C})^n$ be embedded diagonally into $N := \mathrm{M}_n(\mathrm{M}_n(\mathbb{C}))$, denote $A := M' \cong \ell_\infty(n)$ (the commutant of M in N) and let $\phi : N \to M$ be the projection $\phi([x_{ij}]) = \oplus_{j=1}^n x_{jj}$. With (e_{ij}) the usual matrix units in $\mathrm{M}_n(\mathbb{C})$, let y be the unitary element in N defined by*

$$y = [e_{ji}] \quad \textit{(that is, the entry on the position } (i,j) \textit{ is } e_{ji}). \tag{5.2}$$

For each $m \in \mathbb{N}$ denote by $d_{m,n}$ the distance of $\phi(y)$ to the set $S_m := \{\psi(y) : \psi \in \mathrm{EUCP}_M^m(N)\}$, where $\mathrm{EUCP}_M^m(N)$ is the set of all elementary unital completely positive M-bimodule maps of the form

$$\psi(x) = \sum_{k=1}^m a_k^* x a_k, \quad \textit{where } a_k \in A \textit{ and } \sum_{k=1}^m a_k^* a_k = 1. \tag{5.3}$$

Then for each fixed m

$$\lim_{n\to\infty} d_{m,n} = 1.$$

Proof. Given ψ of the form (5.3), noting that each a_k is a diagonal matrix, $a_k = \mathrm{diag}(\alpha_{kj}1)$ ($\alpha_{kj} \in \mathbb{C}$), we compute that the matrix $\psi(y)$ is

$$\psi(y) = \left[\sum_{k=1}^m \overline{\alpha}_{ki} e_{ji} \alpha_{kj}\right].$$

Thus, denoting for each i by $\alpha_i \in \mathbb{C}^m$ the vector $\alpha_i = (\alpha_{1i}, \ldots, \alpha_{mi})$, we have

$$\delta := \|\psi(y) - \phi(y)\| = \|\,[(\langle \alpha_i, \alpha_j\rangle - \delta_{ij})e_{ji}]\,\|. \tag{5.4}$$

From $\psi(1) = 1$ we compute that $\langle \alpha_i, \alpha_i\rangle = 1$ for all $i = 1, \ldots, n$. From (5.4) we have that $|\langle \alpha_i, \alpha_j\rangle| \leq \delta$ if $i \neq j$, hence

$$\|\alpha_i - \alpha_j\|^2 = 2(1 - \mathrm{Re}\langle \alpha_i, \alpha_j\rangle) \geq 2(1-\delta) =: r^2, \text{ if } i \neq j,\ i,j \in \{1, \ldots, n\}. \tag{5.5}$$

This implies that any two of the n balls with centers α_i and radius $r/2$ in $\mathbb{C}^m$ have at most one common point. Since all these balls are contained in the ball of $\mathbb{C}^m$ with center 0 and radius $1 + \frac{r}{2}$, we conclude by considering the Euclidean volume that $n(\frac{r}{2})^{2m+1} \leq (1 + \frac{r}{2})^{2m+1}$, hence from (5.5)

$$\delta = 1 - \frac{r^2}{2} \geq 1 - 2\left(n^{\frac{1}{2m+1}} - 1\right)^{-2}.$$

Since this holds for all ψ, $d_{m,n} \geq 1 - 2(n^{\frac{1}{2m+1}} - 1)^{-2}$, hence (since $d_{m,n} \leq 1$) $d_{m,n}$ necessarily tends to 1 as $n \to \infty$. □

For notational simplicity we shall consider now only a special kind of map of the form (5.1). Given a (strictly) increasing sequence (n_i) in $\mathbb{N}$, let $\mathcal{H} = \oplus_{i=1}^{\infty}(\mathbb{C}^{n_i})^{n_i}$ and consider the diagonal unital embeddings

$$B := \bigoplus_{i=1}^{\infty} \mathrm{M}_{n_i}(\mathbb{C})^{n_i} \subseteq \bigoplus_{i=1}^{\infty} N_i \subseteq \mathrm{B}(\mathcal{H}), \text{ where } N_i = \mathrm{M}_{n_i}(\mathrm{M}_{n_i}(\mathbb{C})). \tag{5.6}$$

For each i let $y_i \in N_i := \mathrm{M}_{n_i}(\mathrm{M}_{n_i}(\mathbb{C}))$ be the unitary defined by (5.2), so that

$$y := \oplus_{i=1}^{\infty} y_i \tag{5.7}$$

is a unitary in $\mathrm{B}(\mathcal{H})$. Finally, let ϕ be the projection of $\mathrm{B}(\mathcal{H})$ onto B, so that

$$\phi(x) = \sum_{i=1}^{\infty} \sum_{k=1}^{n_i} p_k^i x p_k^i \quad (x \in \mathrm{B}(\mathcal{H})), \tag{5.8}$$

where p_k^i is the unit of the kth summand $\mathrm{M}_{n_i}(\mathbb{C})$ of $\mathrm{M}_{n_i}(\mathbb{C})^{n_i}$ inside N_i (a projection in $\mathrm{B}(\mathcal{H})$). Note that

$$B = \mathrm{im}\phi \quad \text{and} \quad A := B' = \bigoplus_{i=1}^{\infty} \bigoplus_{k=1}^{n_i} \mathbb{C} p_k^i = \text{the center of } B.$$

Proposition 5.2. *The inequality $\|\phi(y) - \psi(y)\| \geq 1$ holds for every unital completely positive map ψ on $\mathrm{B}(\mathcal{H})$ of the form*

$$\psi(x) = \sum_{j=1}^{m} a_j^* x a_j, \text{ where } a_j \in A \text{ and } m \in \mathbb{N}. \tag{5.9}$$

Further, ϕ can not be approximated pointwise by elementary complete contractions of the form $x \mapsto \sum_{j=1}^{m} a_j x b_j$, where all a_j or all b_j are in A.

Proof. The first statement follows from Lemma 5.1 by considering the restrictions of ϕ to subalgebras N_i. Then the second statement follows by noting that the estimate (3.1) would allow us to replace completely contractive elementary approximands by unital completely positive elementary approximands as in the proof of Proposition 3.4. □

It can be shown by methods of [20] or [14] that all completely contractive elementary left B-module homomorphisms on $\mathrm{B}(\mathcal{H})$ are of the form $x \mapsto \sum_j a_j x b_j$, where $a_j \in B' = A$ and $b_j \in \mathrm{B}(\mathcal{H})$, hence by Proposition 5.2 the map ϕ can not be approximated pointwise in norm by such homomorphisms.

The question of whether the conditional expectation from $\mathrm{B}(\mathcal{H})$ onto an atomic maximal abelian self-adjoint subalgebra A_0 of $\mathrm{B}(\mathcal{H})$ can be approximated pointwise in norm by completely contractive A_0-bimodule homomorphisms, turns out to be equivalent to the well-known paving problem (hence, to the Kadison–Singer problem), but we shall not present a proof here.

Lemma 5.3. *Let $E : \mathrm{B}(\mathcal{H}) \to A$ be the diagonal projection, θ a unital completely positive map on $\mathrm{B}(\mathcal{H})$ of the form*

$$\theta(x) = \sum_{j=1}^{m} b_j^* x b_j, \quad \text{where } b_j \in \mathrm{B}(\mathcal{H}) \text{ and } m \in \mathbb{N},$$

set $b := (b_1, \ldots, b_m)^T$ and $a := (E(b_1), \ldots, E(b_m))^T =: (a_1, \ldots, a_m)^T$. If $\|\theta(y) - \phi(y)\| \leq 1/4$ (where y and ϕ are as above, defined by (5.7) *and* (5.8)*), then $\|a-b\| > 1/8$.*

Proof. Assume the contrary, that $\|a - b\| \leq 1/8$ and define a map ψ on $\mathrm{B}(\mathcal{H})$ by (5.9) with $a_j = E(b_j)$. By the Schwarz inequality for completely positive maps

$$a^*a = \sum_{j=1}^{m} E(b_j)^* E(b_j) \leq E\Big(\sum_{j=1}^{m} b_j^* b_j\Big) = 1.$$

It follows (since $\theta(y) = b^*yb$ and $\psi(y) = a^*ya$)

$$\begin{aligned} \|\phi(y) - \psi(y)\| &\leq \|\phi(y) - \theta(y)\| + \|b^*yb - a^*ya\| \\ &\leq \tfrac{1}{4} + 2\|b - a\| \leq \tfrac{1}{2}. \end{aligned} \tag{5.10}$$

Further, $\|1 - a^*a\| = \|b^*b - a^*a\| \leq 2\|b - a\| \leq 1/4$. Define a unital completely positive map ψ_0 on $\mathrm{B}(\mathcal{H})$ by $\psi_0(x) = \psi(x) + \sqrt{1 - a^*a}\, x \sqrt{1 - a^*a}$. Then $\|\psi_0 - \psi\|_{\mathrm{cb}} \leq \|1 - a^*a\| \leq 1/4$, hence (5.10) implies that $\|\phi(y) - \psi_0(y)\| \leq 3/4$, which contradicts Proposition 5.2. □

Proposition 5.4. *The map ϕ on $\mathrm{B}(\mathcal{H})$ defined by* (5.8) *cannot be approximated pointwise by any sequence of elementary complete contractions.*

Proof. Suppose the contrary, that a sequence (θ_k) of elementary complete contractions on $\mathrm{B}(\mathcal{H})$ converges pointwise to ϕ. By a simple adjustment of coefficients (as in the proof of Proposition 3.4) we may assume that the maps θ_k are unital and completely positive of the form $\theta_k(x) = \sum_{j=1}^{m_k} b_{k,j}^* x b_{k,j}$, hence the columns $b_k := (b_{k,1}, \ldots, b_{k,m_k})^T \in \mathrm{B}(\mathcal{H})^{m_k}$ satisfy $b_k^* b_k = 1$.

Let $\mathbb{N} = \cup_{j=1}^{\infty} \mathbb{N}_j$ be a partition of $\mathbb{N}$ into a sequence of infinite subsets $\mathbb{N}_j$ and for each j let

$$P_j := \sum_{i \in \mathbb{N}_j} p^i,$$

where for each i the projection $p^i \in \mathrm{B}(\mathcal{H})$ is the unit of N_i (N_i is defined in (5.6)). Then $\|\phi(x) - P_j\theta_k(x)P_j\| = \|P_j(\phi(x) - \theta_k(x))P_j\| \stackrel{k\to\infty}{\longrightarrow} 0$ for each $x \in P_j\mathrm{B}(\mathcal{H})P_j$. If $E : \mathrm{B}(\mathcal{H}) \to A$ is the diagonal projection, then $E|P_j\mathrm{B}(\mathcal{H})P_j$ is the diagonal projection onto AP_j and $E(P_jxP_j) = E(x)P_j$ since $P_j \in A$. Applying Lemma 5.3 to the restriction $\phi|P_1\mathrm{B}(\mathcal{H})P_1$ (with P_1y instead of the unitary y used in the lemma) it follows that

$$\|P_1b_kP_1 - E(b_k)P_1\| > \frac{1}{8} \tag{5.11}$$

for all large enough k, say, for $k \geq k_1$. Now recall that (since $A = B'$ is abelian, hence equal to the weak* closure of a union of an increasing net of finite-dimensional subalgebras) E (= the composition of the conditional expectation ϕ : $\mathrm{B}(\mathcal{H}) \to B$ followed by the central trace $B \to A$) can be approximated in the point-weak* topology by convex combinations of maps of the form $x \mapsto uxu^*$, where $u \in B$ are unitary [12, p. 523 and p. 571]. So, it follows from (5.11) that there exists a unitary $u_1 \in P_1 B$ such that $\|P_1 b_{k_1} P_1 - u_1 b_{k_1} u_1^* P_1\| > 1/8$, that is

$$\|P_1(b_{k_1} u_1 - u_1 b_{k_1})P_1\| > \frac{1}{8}.$$

Similarly, applying Lemma 5.3 to the restriction $\phi|P_2\mathrm{B}(\mathcal{H})P_2$, there exists $k_2 > k_1$ such that $\|P_2 b_k P_2 - E(b_k)P_2\| > 1/8$ for all $k \geq k_2$ and consequently there exists a unitary $u_2 \in P_2 B$ such that

$$\|P_2(b_{k_2} u_2 - u_2 b_{k_2})P_2\| > \frac{1}{8}.$$

Continuing in this way, we find a subsequence (b_{k_j}) of (b_k) and a sequence of unitary elements $u_j \in P_j B$ such that

$$\|P_j(b_{k_j} u_j - u_j b_{k_j})P_j\| > \frac{1}{8}.$$

Then $u := \sum_{j=1}^n u_j$ is a unitary in B and for each i

$$\begin{aligned}\|b_{k_i} u - u b_{k_i}\| &\geq \|P_i(b_{k_i} u - u b_{k_i})P_i\| \\ &= \|P_i(b_{k_i} u P_i - P_i u b_{k_i})P_i\| \\ &= \|P_i(b_{k_i} u_i - u_i b_{k_i})P_i\| > \tfrac{1}{8}.\end{aligned}$$

Since

$$\begin{aligned}\|b_{k_i} u - u b_{k_i}\|^2 &= \|(u^* b_{k_i}^* - b_{k_i}^* u^*)(b_{k_i} u - u b_{k_i})\| \\ &= \|2 - b_{k_i}^* u^* b_{k_i} u - u^* b_{k_i}^* u b_{ki}\| \\ &\leq \|(u - b_{k_i}^* u b_{k_i})^* u\| + \|u^*(u - b_{k_i}^* u b_{k_i})\| \\ &= 2\|u - b_{k_i}^* u b_{k_i}\|,\end{aligned}$$

it follows that

$$\|u - b_{k_i}^* u b_{k_i}\| \geq \frac{1}{2 \cdot 8^2}.$$

But this is a contradiction since $b_{k_i}^* u b_{k_i}$ must converge to $\phi(u) = u$ as $i \to \infty$ (because $u \in B$). □

Elliott [9] proved that pointwise convergence to the identity of a sequence of completely positive maps on a W^*-algebra implies uniform convergence. Our initial proof of Proposition 5.4 was based on techniques of [9]. Later we have found a simpler proof presented above, for which the inspiration of constructing the appropriate unitary u comes from [11].

References

[1] P. Ara and M. Mathieu, *On the central Haagerup tensor product,* Proc. Edinburgh Math. Soc. **37** (1994), 161–174.

[2] P. Ara and M. Mathieu, *Local multipliers of C^*-algebras,* Springer Monographs in Math., Springer-Verlag, Berlin, 2003.

[3] R.J. Archbold, D.W.B. Somerset and R.M. Timoney, *On the central Haagerup tensor product and completely bounded mappings of a C^*-algebra*, J. Funct. Anal. **226** (2005), 406–428.

[4] W.B. Arveson, *Notes on extensions of C^*-algebras*, Duke. Math. J. **44** (1977), 329–355.

[5] D.P. Blecher and C. Le Merdy, *Operator algebras and their modules,* L.M.S. Monographs, New Series **30**, Clarendon Press, Oxford, 2004.

[6] A. Chatterjee and R.R. Smith, *The central Haagerup tensor product and maps between von Neumann algebras,* J. Funct. Anal. **112** (1993), 97–120.

[7] E.G. Effros and R. Exel, *On multilinear double commutation theorems,* Operator algebras and appl., Vol. 1, LMS Lecture Note Series, Vol. 135, Cambridge Univ. Press, Cambridge, 1988, pp. 81–94.

[8] E.G. Effros and A. Kishimoto, *Module maps and Hochschild–Johnson cohomology,* Indiana Univ. Math. J. **36** (1987), 257–276.

[9] G.A. Elliott, *On the convergence of a sequence of completely positive maps to the identity,* J. Austral. Math. Soc. Ser. A **68** (2000), 340–348.

[10] J. Glimm, *A Stone-Weierstrass theorem for C^*-algebras,* Ann. of Math. **72** (1960), 216–244.

[11] B.E. Johnson and S.K. Parrott, *Operators commuting with a von Neumann algebra modulo the set of compact operators,* J. Funct. Anal. **11** (1972), 39–61.

[12] R.V. Kadison and J.R. Ringrose, *Fundamentals of the theory of operator algebras, Vols.* 1, 2, Academic Press, London, 1983, 1986.

[13] B. Magajna, *A transitivity theorem for algebras of elementary operators,* Proc. Amer. Math. Soc. **118** (1993), 119–127.

[14] B. Magajna, *The Haagerup norm on the tensor product of operator modules,* J. Funct. Anal. **129** (1995), 325–348.

[15] B. Magajna, *A transitivity problem for completely bounded mappings,* Houston J. Math. **23** (1997), 109–120.

[16] B. Magajna, *Pointwise approximation by elementary complete contractions,* Proc. Amer. Math. Soc. **137** (2009), 2375–2385.

[17] B. Magajna, *Uniform approximation by elementary operators,* Proc. Edinburgh Math. Soc. **52** (2009), 731–749.

[18] V.I. Paulsen, *Completely bounded maps and operator algebras,* Cambridge Studies in Advanced Mathematics **78**, Cambridge University Press, Cambridge, 2002.

[19] H. Pfitzner, *Weak compactness in the dual of a C^*-algebra is determined commutatively,* Math. Ann. **298** (1994), 349–371.

[20] R.R. Smith, Completely bounded module maps and the Haagerup tensor product, *J. Funct. Anal.* **102** (1991), 156–175.

[21] D.W.B. Somerset, *The central Haagerup tensor product of a C^*-algebra*, J. Operator Theory **39** (1998), 113–121.

[22] M. Takesaki, *Theory of operator algebras I,* Springer-Verlag, New York, 1979.

[23] D.V. Voiculescu, *A non-commutative Weyl–von Neumann theorem,* Rev. Roum. Math. Pures Appl. **21** (1976), 97–113.

Bojan Magajna
Department of Mathematics
University of Ljubljana
Jadranska 21
1000 Ljubljana, Slovenia
e-mail: `bojan.magajna@fmf.uni-lj.si`

Some Not-quite-elementary Operators

Peter Rosenthal

Abstract. Necessary and sufficient conditions are given that an operator equation of the form $AXB+CYD=Z$ have a solution (X,Y) for every operator Z.

Mathematics Subject Classification (2000). Primary 47A62; Secondary 47B49, 15A24.

Keywords. Operator equations, matrix equations.

Let $\mathcal{B}(\mathcal{H})$ denote the algebra of all bounded linear operators on the complex Hilbert space $\mathcal{H}$. An *elementary operator* is a mapping of the form

$$X \to \sum_{i=1}^{n} A_i X B_i,$$

where the $\{A_i\}$ and the $\{B_i\}$ are in $\mathcal{B}(\mathcal{H})$. The "not-quite-elementary operators" of the title are mappings from the direct sum of n copies of $\mathcal{B}(\mathcal{H})$ into $\mathcal{B}(\mathcal{H})$ of the form

$$(X_1, X_2, \ldots, X_n) \to \sum_{i=1}^{n} A_i X_i B_i$$

where the $\{A_i\}$ and the $\{B_i\}$ are in $\mathcal{B}(\mathcal{H})$. Such operators appear to have been first studied in [1]. This expository paper discusses some of the results of [1] and presents a more direct proof of what might be regarded as the main theorem of [1]. (See Theorem 2 below.)

Elementary operators originated from the study of the Sylvester–Rosenblum equation $AX - XB = Z$. (See [2] for a discussion of the many known results about this equation and of the wide variety of applications of those results.) Don Hadwin, Eric Nordgren, and I got interested in the study of "not-quite-elementary operators" because of our observation about the operator equation $AX - YB = Z$ in the finite-dimensional case. The surprising theorem is the following:

Theorem 1. [1] *For square complex matrices A and B, the mapping*

$$(X, Y) \to AX - YB$$

is onto if and only if at least one of the mappings $X \to AX$ or $Y \to YB$ is onto.

R.E. Curto and M. Mathieu (eds.), *Elementary Operators and Their Applications*, Operator Theory: Advances and Applications 212, DOI 10.1007/978-3-0348-0037-2_4,

Proof. Suppose that the map $(X,Y) \to AX - YB$ is onto. If the map $Y \to YB$ is not onto, then B has nullspace. The proof will be completed by showing that the map $X \to AX$ is onto in this case. Let f be a vector other than 0 such that $Bf = 0$. If $X \to AX$ is not onto, then A itself is not onto. Choose a g that is not in the range of A and pick any matrix Z such that $Zf = g$. Then Z cannot be of the form $AX - YB$, since $(AX - YB)f = AXf$, which is not Zf since g is not in the range of A. □

The same theorem holds in the infinite-dimensional case. In fact, a more general result was established in [1]. We rely on the following basic facts about operators. Recall that an operator is bounded below if and only if it has a left inverse. We will also use the obvious fact that an operator has a right inverse if and only if its adjoint has a left inverse.

If the operator A is not bounded below, then, by definition, there exists a sequence $\{f_n\}$ of unit vectors such that $\{Af_n\} \to 0$. It follows by an elementary computation that goes back to Dixmier [3] (which can also be found in [4] and in [5]) that the sequence $\{f_n\}$ can be taken to be orthonormal. These elementary facts are all we need to prove the general theorem in the case of two summands.

If the mapping $(X_1, X_2) \to A_1X_1B_1 + A_2X_2B_2$ is onto, then clearly $(A_1\ A_2)$ takes $\mathcal{H} \oplus \mathcal{H}$ onto $\mathcal{H}$ and taking adjoints shows that the same is true for $(B_1^*\ B_2^*)$. The theorem is the following:

Theorem 2. [1] *The mapping* $(X_1, X_2) \to A_1X_1B_1 + A_2X_2B_2$ *is onto if and only if*

a) *both* $(A_1\ A_2)$ *and* $(B_1^*\ B_2^*)$ *take* $\mathcal{H} \oplus \mathcal{H}$ *onto* $\mathcal{H}$, *and*
b) *both of the operators in at least one of the following pairs are onto:* $\{A_1, B_1^*\}$, $\{A_2, B_2^*\}$, $\{A_1, A_2\}$, *and* $\{B_1^*, B_2^*\}$.

Proof. First suppose that the map defined by $\Phi(X_1, X_2) = A_1X_1B_1 + A_2X_2B_2$ is onto. Then a) clearly holds, so we must establish b). If neither of the elements of the set $\{B_1^*, A_2\}$ is onto, then their adjoints are not bounded below, so there exist orthonormal sequences $\{f_n\}$ and $\{g_n\}$ such that $\{B_1f_n\} \to 0$ and $\{A_2^*g_n\} \to 0$. Then, for every X_1 and X_2,

$$((A_1X_1B_1 + A_2X_2B_2)f_n, g_n) = (A_1X_1B_1f_n, g_n) + (X_2B_2f_n, A_2^*g_n)$$

converges to 0 as n approaches infinity. On the other hand, since $\{f_n\}$ and $\{g_n\}$ are orthonormal sequences, there is an operator Z such that $Zf_n = g_n$ for every n. Such an operator cannot be in the range of Φ since $(Zf_n, g_n) = 1$ for all n. Hence at least one of the operators in the set $\{B_1^*, A_2\}$ is onto.

In exactly the same way, it can be established that at least one of the operators in the set $\{B_2^*, A_1\}$ is onto. To get the conclusion of part b) of the theorem we now simply note the following: if at least one of the operators in each of the sets $\{B_1^*, A_2\}$ and $\{B_2^*, A_1\}$ is onto, then both of the operators in at least one of the sets $\{A_1, B_1^*\}$, $\{A_2, B_2^*\}$, $\{A_1, A_2\}$, $\{B_1^*, B_2^*\}$ are onto.

For the converse, it must be shown that conditions a) and b) imply that Φ is onto. Suppose then that a) holds. If both A_1 and B_1^* are onto, then there exist operators C_1 and D_1 such that $A_1C_1 = I$ and $D_1B_1 = I$. Hence, for any operator X, it follows that $A_1(C_1XD_1)B_1 = X$, so Φ is onto. The same proof works if both A_2 and B_2^* are onto.

Suppose now that both of A_1 and A_2 are onto. Then there exist operators E_1 and E_2 such that $A_1E_1 = I$ and $A_2E_2 = I$. Now, $(B_1^*\, B_2^*)$ is onto, so there exist operators F_1 and F_2 such that $B_1^*F_1 + B_2^*F_2 = I$. Given any Z then, let $X_1 = E_1ZF_1^*$ and $X_2 = E_2ZF_2^*$. It follows that

$$\begin{aligned}\Phi(X_1, X_2) &= A_1(E_1ZF_1^*)B_1 + A_2(E_2ZF_2^*)B_2 \\ &= ZF_1^*B_1 + ZF_2^*B_2 \\ &= Z(F_1^*B_1 + F_2^*B_2) \\ &= Z.\end{aligned}$$

Thus in this case Φ is onto. The case where B_1^* and B_2^* are both onto is established in exactly the same way after taking adjoints. □

The fact that Theorem 1 holds for operators on Hilbert space is a special case of Theorem 2.

Corollary 3. [1] *For A and B operators on Hilbert space, the mapping*

$$(X, Y) \to AX - YB$$

is onto if and only if at least one of the mappings $X \to AX$ or $Y \to YB$ is onto.

Proof. In the notation of Theorem 2, the corollary is the special case where $A_1 = A$, $B_1 = I$, $A_2 = -I$, and $B_2 = B$. Thus condition a) of Theorem 2 is automatically satisfied. Moreover, B_1^* and A_2 are clearly onto. Therefore b) reduces to the condition that at least one of A and B^* be onto, which is the conclusion of the corollary. □

In [1] there are some partial results obtained on mappings of the form $(X_1, X_2, \ldots, X_n) \to \sum_{i=1}^n A_iX_iB_i$ where $n \geq 3$. It is shown that the question of when Φ is onto can be reduced to the case where the coefficients $\{A_i\}$ and $\{B_i\}$ are projections. A necessary condition that a Φ with projection coefficients be onto is obtained. There is a sufficient condition established in the special case where each of the $\{A_i\}$ and the $\{B_i\}$ consists of commuting families of projections. Moreover, in the case $n = 2$, Theorem 2 above is also proven for Φ restricted to the algebra of compact operators and to the Schatten p-classes. However, necessary and sufficient conditions that such a mapping be onto are only given in the case $n = 2$. Thus there is more work to be done on such mappings.

Another question left open in [1] is whether the corollary holds for operators on Banach spaces. More generally, the paper [1] does not contain any results on Banach spaces that are not Hilbert spaces.

References

[1] D. Hadwin, E. Nordgren, and P. Rosenthal, *On the operator equation* $AXB+CYD = Z$, Oper. Matrices, **1** (2007), no. 2, 199–207.

[2] R. Bhatia and P. Rosenthal, *How and Why to Solve the Operator Equation* $AX - XB = Y$, Bull. London Math. Soc. **29** (1997), 1–21.

[3] J. Dixmier, *Etude sur les variétés et les opérateurs de Julia, avec quelques applications*, Bull. Soc. Math. France **77** (1949), 11–101.

[4] P.A. Fillmore, J.G. Stampfli, and J.P Williams, *On the essential numerical range, the essential spectrum, and a problem of Halmos*, Acta Sci. Math. (Szeged) **33** (1972), 179–192.

[5] E. Nordgren and P. Rosenthal, *Boundary Values of Berezin Symbols*, Operator Theory: Advances and Applications, Vol. **73**, Birkhäuser Verlag Basel/Switzerland, 1994.

Peter Rosenthal
Department of Mathematics
University of Toronto
Toronto, Ontario M5S 3G3
Canada
e-mail: `rosent@math.toronto.edu`

Topological Radicals, II. Applications to Spectral Theory of Multiplication Operators

V.S. Shulman and Yu.V. Turovskii

To Wojciech Wojtyński with gratitude for his wonderful results and stimulating questions.

Abstract. We develop the tensor spectral radius technique and the theory of the tensor Jacobson radical. Based on them we obtain several results on spectra of multiplication operators on Banach bimodules and indicate some applications to the spectral theory of elementary and multiplication operators on Banach algebras and modules with various compactness properties.

Mathematics Subject Classification (2000). Primary 46H15, 46H25; Secondary 16Nxx.

Keywords. Tensor spectral radius, joint spectral radius, tensor Jacobson radical, hypocompact radical, hypofinite radical, elementary operator, multiplication operator, semicompact operator, semifinite operator, hypocompact algebra, hypofinite algebra.

1. Introduction

The localization of spectrum of an elementary operator in terms of spectra of its coefficients is one of the most popular subjects in the theory of elementary operators. The strongest results in this area were obtained for operators with commutative coefficient families because this allows one to use the theory of joint spectra (see [11]).

Here we consider the less restrictive conditions than commutativity. For instance it is not known to us whether an operator $Tx = \sum_{k=1}^{n} a_k x b_k$ on a Banach algebra A is quasinilpotent if $a_1, \ldots, a_n$ belong to a radical closed subalgebra of A. However, if all a_i are compact operators, the answer is positive (see for example

---

This paper is in final form and no version of it will be submitted for publication elsewhere.

R.E. Curto and M. Mathieu (eds.), *Elementary Operators and Their Applications*, Operator Theory: Advances and Applications 212, DOI 10.1007/978-3-0348-0037-2_5, © Springer Basel AG 2011

[28, Lemma 5.10]) and may be obtained by using the joint spectral radius technique. We consider multiplication operators of more general type than elementary ones as well as more general classes of coefficient algebras than algebras of compact operators. As a main technical tool we present the theory of tensor spectral radius initiated in [29] in the framework of the general theory of topological radicals. Based on it we obtain several results on spectra of multiplication operators on Banach bimodules and indicate their applications to spectral theory of elementary operators on Banach algebras with various compactness properties.

Recall that an element a of a normed algebra A is called *compact* if the elementary operator $x \longmapsto axa$ on A is compact. The reason for such a definition is a well-known theorem of Vala [32] which states that *a bounded operator on a Banach space X is compact iff it is a compact element of the algebra $\mathcal{B}(X)$ of all bounded operators on X.*

If all elements $a \in A$ are compact then A is called *compact.* If, more strongly, for all $a, b \in A$, the operator $x \longmapsto axb$ on A is compact then A is called *bicompact.* A less restrictive condition is that A is generated as a normed algebra by the semigroup of all its compact elements. The most wide class of algebras of this kind is the class of hypocompact algebras. A normed algebra A is called *hypocompact* if each non-zero quotient of A by a closed ideal has a non-zero compact element. One may realize a hypocompact algebra as a result of a transfinite sequence of extensions of bicompact algebras. This class has some resemblance with the class of GCR-algebras in the C^*-algebras. Note for example that *the image of each strictly irreducible representation of a hypocompact Banach algebra contains a non-zero finite rank operator.*

We show that elementary operators on hypocompact Banach algebras commutative modulo the Jacobson radical are *spectrally computable*, that is

$$\sigma(T+S) \subset \sigma(T) + \sigma(S) \text{ and } \sigma(TS) \subset \sigma(T)\sigma(S)$$

for all elementary operators T, S. Moreover, if all operators $L_a - R_a$ are quasinilpotent on A (we call such algebras *Engel*) then the spectra of elementary operators satisfy the inclusion

$$\sigma\left(\sum_k L_{a_k} R_{b_k}\right) \subset \sigma\left(\sum_k a_k b_k\right).$$

Among other applications we mention results on the structure of closed ideals in a radical compact Banach algebra. We prove that if such an algebra is infinite dimensional then it has infinite chains of ideals. As a consequence, we get that there is an infinite chain of closed operator ideals in the sense of Pietsch [20] intermediate between the ideals of approximable and compact operators.

In the last section we consider the applications of the theory to spectral subspaces of multiplication operators. In 1978 Wojtyński, working on the problem of the existence of a closed two-sided ideal in a radical Banach algebra, proved the following result on linear operator equations with compact coefficients.

Lemma 1.1. [36] *Let all coefficients a, b, a_i, b_i of the linear operator equation*

$$ax + xb + \sum_{i=1}^{n} a_i x b_i = \lambda x \tag{1.1}$$

be compact operators on a Banach space X. If $\lambda \neq 0$ then each bounded solution x of (1.1) *is a nuclear operator.*

The presence of nuclear operators gives a possibility to use trace for proving the quasinilpotence of some multiplication operators. Using this, Wojtyński proved in [36] that *every radical Banach algebra having non-zero compact elements is not topologically simple* (if dimension of the algebra is larger than 1). In [35] he applied the same argument to Banach Lie algebras and proved that *if all adjoint operators of a Banach Lie algebra $\mathfrak{L}$ are compact and quasinilpotent, then $\mathfrak{L}$ has a non-trivial closed Lie ideal.* Both results are now obtained in a more general setting with using another technique [31, 28], but Wojtyński's approach itself is interesting and still helpful.

Several years after [36], Fong and Radjavi [14] considered a more general class of equations

$$\sum a_i x b_i = \lambda x, \tag{1.2}$$

where the sum is finite and for each i at least one of operators a_i, b_i is compact. They worked only in the case of Hilbert space operators but proved much more, namely that all solutions of (1.2) belong to each Schatten class C_p, $p > 0$. On the other hand, they showed that solutions of (1.2) are not necessarily finite rank operators: each operator x whose singular numbers decrease more quickly than every geometric progression is a solution of an equation of the form (1.2).

We will show here that the main results of Wojtyński and of Fong and Radjavi extend to multiplication operators with infinite number of summands. Furthermore, we will see that not only eigenspaces with non-zero eigenvalues consist of nuclear operators but that the same holds for spectral subspaces corresponding to components of spectra non-containing 0 (or stronger, for all invariant subspaces on which the operator is surjective). Moreover, the ideal of nuclear operator here can be changed by any quasi-Banach ideal in the case of elementary operators (i.e., when the number of summands is finite).

We prove also that the results extend to "integral multiplication operators". In particular, if an operator x satisfies the condition

$$\int_{\alpha}^{\beta} (a(t)xu(t) + v(t)xb(t))d\mu = \lambda x \quad \text{with} \quad \lambda \neq 0,$$

where $a(t)$ and $b(t)$ are continuous operator-valued functions, $u(t)$ and $v(t)$ are continuous compact operator-valued functions, then x is nuclear.

We also extend the results to systems of equations. A simple example is the following: Let $x_1, \dots, x_n$ satisfy a system of equations

$$\sum_{k=1}^{n} a_{ik} x_k b_{ik} = \lambda x_i, \ i = 1, \dots, n, \lambda \neq 0.$$

If for each pair (i, k) at least one of operators a_{ik}, b_{ik} is compact then all x_i are nuclear (moreover belong to each quasi-Banach operator ideal of $\mathcal{B}(X)$). Such systems of equations arise, for example, in the study of subgraded Lie algebras [16].

Apart of tensor radical technique our approach is based on a general result (Theorem 6.5) which is not restricted by multiplication operators but deals with bounded operators on an ordered pair of Banach spaces.

We would like to express our heartfelt gratitude to Niels Grønbæk for a very helpful discussion of the results of the paper [34] and to the referee for his patient and attentive reading of the manuscript and for numerous useful suggestions.

2. Preliminary results

2.1. Notation

All spaces are assumed to be complex. If a normed algebra A is not unital, denote by A^1 the normed algebra obtained by adjoining the identity element to A, and if A is already unital, let $A^1 = A$. We denote by $\widehat{A}$ the completion of A. The term *ideal* always means a two-sided ideal. If A is a normed algebra and I is an ideal of A then the term $A/\overline{I}$ always denotes the quotient of A by the closure of I *in the norm of* A (even if I is supplied with its own norm). It is convenient to write $a/\overline{I}$ for $a + \overline{I} \in A/\overline{I}$. Also, by a *quotient* of a normed algebra A we always mean any quotient of A by a closed ideal.

Let A be a normed algebra. A norm $\|\cdot\|_A$ on A is called an *algebra* (or *submultiplicative*) *norm* if

$$\|ab\|_A \leq \|a\|_A \|b\|_A$$

for all $a, b \in A$. If A has another norm (or seminorm), say $\|\cdot\|$, we write $(A, \|\cdot\|)$ to indicate that A is considered with respect to $\|\cdot\|$. The norm $\|\cdot\|$ is *equivalent* to $\|\cdot\|_A$ on A if there are constants $s, t > 0$ such that

$$s \|\cdot\| \leq \|\cdot\|_A \leq t \|\cdot\|$$

on A. Assume now that A is unital. Then $\|\cdot\|_A$ is called *unital* if $\|1\|_A = 1$. It is well known that every algebra norm on A is equivalent to a unital one.

2.2. Quasinilpotents and the radical modulo an ideal

An element a of a normed algebra A is called *quasinilpotent* if

$$\inf_n \|a^n\|^{1/n} = 0.$$

Let $Q(A)$ denote the set of all quasinilpotent elements of A.

Let A be a normed algebra, and let $\operatorname{dist}_A(a, E)$, or simply $\operatorname{dist}(a, E)$, denote the distance from $a \in A$ to $E \subset A$, that is

$$\operatorname{dist}(a, E) = \inf \{\|a - b\| : b \in E\}.$$

Let $Q_E(A)$ be the set of all elements $a \in A$ such that

$$\inf_n \operatorname{dist}_A(a^n, E)^{1/n} = 0.$$

If $E = J$ is an ideal of A, then $\operatorname{dist}_A(a, J)$ is simply a quotient norm of $q(a)$ in the quotient algebra $A/\overline{J}$, where $\overline{J}$ denotes the closure of J in A and $q : A \longrightarrow A/\overline{J}$ is the standard quotient map. In other words, $Q_J(A)$ is the set of all $a \in A$ quasinilpotent modulo J.

Let $\operatorname{rad}(A)$ denote the Jacobson radical of A, and let $\operatorname{rad}_J(A)$ denote the Jacobson radical of A modulo J, that is the preimage in A of the radical of the quotient algebra $A/\overline{J}$. If A is complete, write $\operatorname{Rad}(A)$ instead of $\operatorname{rad}(A)$.

Proposition 2.1. *Let A be a Banach algebra and J be an ideal of A. Then*

(i) *$Q_J(A)$ contains the intersection of all primitive ideals of A containing J* $(= \operatorname{Rad}_J(A))$.

(ii) *If a closed subalgebra $B \subset A$ is such that $J \subset B$ and $B/\overline{J}$ is radical then* $B \subset Q_J(A)$.

(iii) *An element $a \in A$ belongs to $Q_J(A)$ if and only if for each $\varepsilon > 0$ there is $n \in \mathbb{N}$ such that* $\operatorname{dist}(a^m, J) < \varepsilon^m$ *for all $m \geq n$.*

Proof. (i) As is known the Jacobson radical $\operatorname{Rad}(A)$ of a Banach algebra A is the largest ideal consisting of quasinilpotents. Hence $Q_J(A)$ contains the Jacobson radical of A modulo $\overline{J}$. The last, as well known, is the intersection of all primitive ideals containing J.

(ii) Straightforward.

(iii) Follows immediately from the equality $\operatorname{dist}(a, J) = \|a/\overline{J}\|$. □

An algebra is usually said to be *radical* if it is Jacobson radical. The following lemma slightly improves the respective classical result.

Lemma 2.2. *Let A be a Banach algebra, and let I be an ideal of A. If I is radical and $A/\overline{I}$ is radical then A is radical.*

Proof. Let π be a strictly irreducible representation of A on X. We may assume that X is a Banach space and that π is continuous. If the restriction of π to I is non-zero then it is a strictly irreducible representation of I, in contradiction with the assumption that I is radical. Therefore $\pi|_I = 0$ and, by continuity, $\overline{I} \subset \ker \pi$. Hence π defines a strictly irreducible representation of $A/\overline{I}$. Since $A/\overline{I}$ is radical, this means that $\dim X = 1$ and $\pi = 0$. □

2.3. Normed subalgebras and flexible ideals

2.3.1. Spectrum with respect to a Banach subalgebra. Let A, B be normed algebras with norms $\|\cdot\|_A$ and $\|\cdot\|_B$ respectively, and let B be a subalgebra of A. We say that B is a *normed subalgebra* if $\|\cdot\|_A \leq \|\cdot\|_B$ on B. Every complete (with respect to $\|\cdot\|_B$) normed subalgebra B is called a *Banach subalgebra.*

Let $\sigma_A(a)$, or simply $\sigma(a)$, denote the spectrum of $a \in A$ with respect to A^1. Recall that this definition of spectrum coincides with Definition 5.1 in [9] in virtue of [9, Lemma 5.2]. Let $\widehat{\sigma}_A(a)$ denote the polynomially convex hull of $\sigma_A(a)$, and let $\rho_A(a)$, or simply $\rho(a)$, denote the spectral radius of a defined as $\inf_n \|a^n\|_A^{1/n}$.

If A is a Banach algebra then $\rho_A(a) = \sup\{|\lambda| : \lambda \in \sigma_A(a)\}$ (Gelfand's formula), and $\widehat{\sigma}_A(a)$ is received from $\sigma_A(a)$ by filling the holes of $\sigma_A(a)$. The term "clopen" means "closed and open simultaneously".

Proposition 2.3. *Let A be a unital Banach algebra, and let B be a unital Banach subalgebra of A (i.e., the units for A and B coincide). Then*

(i) *$\sigma_A(a) \subset \sigma_B(a)$ for every $a \in B$, and each clopen subset of $\sigma_B(a)$ has a non-void intersection with the polynomial hull $\widehat{\sigma}_A(a)$ of $\sigma_A(a)$.*
(ii) *If $\sigma_B(A)$ is finite or countable then $\sigma_A(a) = \sigma_B(a)$.*

Proof. It is evident that $\sigma_A(a) \subset \sigma_B(a)$. To prove the second statement, suppose that σ_1 is a clopen subset of $\sigma_B(a)$ which doesn't intersect $\widehat{\sigma}_A(a)$. Let p be the corresponding Riesz projection in B,

$$p = \frac{1}{2\pi i} \int_\Gamma (\lambda - a)^{-1} d\lambda,$$

where Γ surrounds σ_1 and doesn't intersect $\widehat{\sigma}_A(a)$. Then $p \neq 0$.

On the other hand, $p = 0$ because it can be regarded as a Riesz projection of a in A and there are no points of $\sigma_A(a)$ inside Γ. The obtained contradiction proves (i).

To show (ii), note that if $\sigma_B(a)$ is countable then $\sigma_A(a)$ is countable hence $\widehat{\sigma}_A(a) = \sigma_A(a)$. Thus each clopen subset of $\sigma_B(a)$ intersects $\sigma_A(a)$. Any point $\lambda \in \sigma_B(a)$ is clearly the intersection of a sequence of clopen subsets of $\sigma_B(a)$. Since all of them intersects $\sigma_A(a)$ we get that $\lambda \in \sigma_A(a)$. Thus $\sigma_B(a) \subset \sigma_A(a)$ and we are done. □

When the subalgebra B is closed in A, the result is related to [22, Theorem 10.18]. The situation is especially simple if B is a (non-necessarily closed) ideal of A.

Remark 2.4. Let I be an ideal of an algebra A and $a \in I$. It is easy to check that if $(a - \lambda) b = 1$ or $b(a - \lambda) = 1$ for $b \in A^1$ and $\lambda \neq 0$, then $b + \lambda^{-1} \in I$. Hence, in virtue of [9, Lemma 5.2], $\{0\} \cup \sigma_I(a) = \{0\} \cup \sigma_A(a)$.

2.3.2. Flexible ideals. Let A, I be normed algebras with norms $\|\cdot\|_A$ and $\|\cdot\|_I$ respectively, and let I be an ideal of A such that

$$\|x\|_A \leq \|x\|_I \text{ and } \|axb\|_I \leq \|a\|_A \|x\|_I \|b\|_A$$

for all $x \in I$ and $a, b \in A^1$. Such an algebra norm $\|\cdot\|_I$ on I is called *flexible* (with respect to $\|\cdot\|_A$ or $(A, \|\cdot\|_A)$, naturally). An ideal having a flexible norm is called a *flexible ideal.* Every ideal I of a normed algebra A with $\|\cdot\|_I = \|\cdot\|_A$ on I is of course flexible.

By definition, a Banach ideal I of a normed algebra A is an ideal which is a Banach subalgebra of A.

Lemma 2.5. *Every Banach ideal of a Banach algebra is flexible with respect to an equivalent algebra norm.*

Proof. Let I be a Banach ideal of a Banach algebra A. By [4, Theorem 2.3], there is $s > 0$ such that $\|axb\|_I \leq s \|a\|_A \|x\|_I \|b\|_A$ for all $x \in I$ and $a, b \in A^1$. Define $\|\cdot\|_I'$ on I by

$$\|x\|_I' = \sup\left\{\|axb\|_I : \|a\|_A, \|b\|_A \leq 1, a, b \in A^1\right\}$$

for every $x \in I$. It is easy to check that $\|\cdot\|_I'$ is an algebra norm on I,

$$\|\cdot\|_I \leq \|\cdot\|_I' \leq s \|\cdot\|_I .$$

and

$$\|axb\|_I' \leq \|a\|_A \|x\|_I' \|b\|_A$$

for every $a, b \in A^1$. As $\|\cdot\|_A \leq \|\cdot\|_I'$ on I, we obtain that I is a flexible ideal with respect to $\|\cdot\|_I'$. □

2.3.3. Completion of normed subalgebras and ideals. If B is a normed subalgebra in a normed algebra A then the "identity" homomorphism $\mathrm{i} : (B, \| \cdot \|_B) \to A$ is continuous and therefore extends by continuity to the homomorphism $\hat{\imath} : \widehat{B} \to \widehat{A}$. The proof of the following result is straightforward and we omit it.

Lemma 2.6. *Let B be a normed subalgebra of a normed algebra A. Then*

(i) *The image $\hat{\imath}(\widehat{B})$ of the completion $\widehat{B}$ of B in the completion $\widehat{A}$ of A is a Banach subalgebra of $\widehat{A}$ with respect to the norm $\|\cdot\|_*$ of the quotient $\widehat{B}/\ker\hat{\imath}$.*

(ii) *If B is a flexible ideal of A then $\hat{\imath}(\widehat{B})$ is a Banach ideal of $\widehat{A}$ and its norm $\|\cdot\|_*$ is flexible.*

When it cannot lead to a misunderstanding, we will write $\widehat{B}^{(A)}$ or, simply, $\widehat{B}^{(\cdot)}$ instead of $\hat{\imath}(\widehat{B})$.

2.3.4. Sums and intersections of Banach ideals. The following extends the class of examples of flexible ideals.

Proposition 2.7. *Let I and J be flexible ideals of a normed algebra A. Then*

(i) *$I \cap J$ is a flexible ideal of A with respect to the norm $\|\cdot\|_{I\cap J} = \max\{\|\cdot\|_I, \|\cdot\|_J\}$.*
(ii) *$I + J$ is a flexible ideal of A with respect to the norm*

$$\|z\|_{I+J} = \inf\{\|x\|_I + \|y\|_J : z = x + y, x \in I, y \in J\}$$

for every $z \in I + J$.
(iii) *$I \cap J$ is a flexible ideal of $I + J$.*
(iv) *If I and J are Banach ideals then $I \cap J$ and $I + J$ are Banach ideals with flexible norms $\|\cdot\|_{I\cap J}$ and $\|\cdot\|_{I+J}$ respectively.*

Proof. It follows from [6, Lemma 2.3.1] that $\|\cdot\|_{I\cap J}$ and $\|\cdot\|_{I+J}$ are norms and (iv) holds if (i) and (ii) hold. So it suffices to show that $\|\cdot\|_{I\cap J}$ and $\|\cdot\|_{I+J}$ are flexible. It is easy to see that $\|\cdot\|_{I\cap J}$ is flexible and $\|\cdot\|_A \le \|\cdot\|_{I+J}$ on $I + J$.

For $z = x + y$ and $z' = x' + y'$ with $x, x' \in I$ and $y, y' \in J$, we obtain that

$$\begin{aligned}
\|zz'\|_{I+J} &\le \|xx' + xy'\|_I + \|yx' + yy'\|_J \le \|xx'\|_I + \|xy'\|_I + \|yx'\|_J + \|yy'\|_J \\
&\le \|x\|_I \|x'\|_I + \|x\|_I \|y'\|_A + \|y\|_J \|x'\|_A + \|y\|_J \|y'\|_J \\
&\le \|x\|_I \|x'\|_I + \|x\|_I \|y'\|_J + \|y\|_J \|x'\|_I + \|y\|_J \|y'\|_J \\
&\le (\|x\|_I + \|y\|_J)(\|x'\|_I + \|y'\|_J)
\end{aligned}$$

and

$$\begin{aligned}
\|azb\|_{I+J} &\le \|axb\|_I + \|ayb\|_J \le \|a\|_A \|x\|_I \|b\|_A + \|a\|_A \|y\|_J \|b\|_A \\
&= \|a\|_A (\|x\|_I + \|y\|_J) \|b\|_A
\end{aligned}$$

for every $a, b \in A^1$. Hence $\|\cdot\|_{I+J}$ is clearly a flexible norm.

(iii) It is clear that $I \cap J$ is an ideal of $I + J$, $\|\cdot\|_{I+J} \le \|\cdot\|_{I\cap J}$ on $I \cap J$, and flexibility of $\|\cdot\|_{I\cap J}$ with respect to $I + J$ follows from one with respect to A. □

In the situation of Proposition 2.7 it is convenient to call $I \cap J$ and $I + J$ with their flexible norms a *flexible intersection* and a *flexible sum* of ideals I and J, respectively.

An important class of examples of flexible ideals may be obtained by using the notion of normed operator ideals [20, 12]. Note that normed operator ideals in [20] are the same as Banach operator ideals in [12], and we prefer the terminology in [12].

Example 2.8. Let $\mathcal{I}$ be a Banach operator ideal. Then $\mathcal{I}(X)$ is a Banach ideal of $\mathcal{B}(X)$ for every Banach space X and its norm is flexible.

2.4. Projective tensor products

2.4.1. Tensor products of normed algebras. Let $A_1 \otimes A_2$ denote the algebraic tensor product of normed algebras A_1 and A_2, and let $A_1 \otimes_\gamma A_2$ denote $(A_1 \otimes A_2, \gamma)$, where γ is the projective crossnorm. Recall that γ is defined by

$$\gamma(c) = \inf\left\{\sum \|a_i\|_{A_1} \|b_i\|_{A_2} : \sum a_i \otimes b_i = c\right\}$$

for every $c \in A_1 \otimes_\gamma A_2$. Then $A = A_1 \otimes_\gamma A_2$ is a normed algebra and γ is its algebra norm. To underline that the projective norm γ is considered in $A_1 \otimes A_2$, we write $\gamma = \gamma_{A_1,A_2}$ or $\gamma = \gamma_A$. We also write $\gamma = \gamma_{\|\cdot\|_{A_1},\|\cdot\|_{A_2}}$ to indicate which norm are considered in A_i.

Let A_1 and A_2 be normed algebras. By $A_1 \widehat{\otimes}_\gamma A_2$, or simply $A_1 \widehat{\otimes} A_2$, we denote the projective tensor product of A_1 and A_2 that is the completion of $A_1 \otimes_\gamma A_2$. By definition, it is a Banach algebra, and clearly it coincides with the projective tensor product of the completions of A_i. The elements of $A_1 \widehat{\otimes} A_2$ can be written in the form

$$c = \sum_{k=1}^{\infty} a_k \otimes b_k \text{ with } \sum_k \|a_k\| \|b_k\| < \infty, \tag{2.1}$$

where $a_k \in A_1, b_k \in A_2$. Moreover, the norm $\|\cdot\| = \gamma(\cdot)$ in $A_1 \widehat{\otimes} A_2$ is given by

$$\|c\| = \inf \sum_k \|a_k\| \|b_k\|,$$

where inf is taken over representations of c in form (2.1).

2.4.2. Tensor products of normed subalgebras and ideals. If B_i is a subalgebra of an algebra A_i for $i = 1, 2$, then $B_1 \otimes_\gamma B_2$ is a subalgebra of $A_1 \otimes_\gamma A_2$ (see [10, Section 3.3.1]). If I_i is an ideal of A_i for $i = 1, 2$, then $I_1 \otimes_\gamma I_2$ is clearly an ideal of $A_1 \otimes_\gamma A_2$.

Proposition 2.9. *Let A_1 and A_2 be normed algebras, and $A = A_1 \otimes_\gamma A_2$. Then*

(i) *If B_i is a normed subalgebra of A_i for $i = 1, 2$, then $B := B_1 \otimes_\gamma B_2$ with $\gamma_B = \gamma_{\|\cdot\|_{B_1},\|\cdot\|_{B_2}}$ is a normed subalgebra of A.*

(ii) *If I_i is a flexible ideal of A_i for $i = 1, 2$, then $I := I_1 \otimes_\gamma I_2$ with $\gamma_I = \gamma_{\|\cdot\|_{I_1},\|\cdot\|_{I_2}}$ is a flexible ideal of A and $\gamma_I(azb) \le \gamma_A(a)\gamma_I(z)\gamma_A(b)$ for every $a, b \in A^1$ and $z \in I$.*

Proof. (i) Indeed, the norm γ_B on B majorizes γ_A (and the equality does not hold in general even if $\|\cdot\|_{B_i} = \|\cdot\|_{A_i}$ on B_i).
(ii) Straightforward. □

The natural embedding i of $B_1 \otimes_\gamma B_2$ into $A_1 \otimes_\gamma A_2$ extends by continuity to a continuous homomorphism $\hat{\imath}$ of $B_1 \widehat{\otimes}_\gamma B_2$ into $A = A_1 \widehat{\otimes}_\gamma A_2$. Let $\hat{\imath}(B_1 \widehat{\otimes}_\gamma B_2)$ be supplied with the norm inherited from the quotient $\left(B_1 \widehat{\otimes}_\gamma B_2\right) / \ker \hat{\imath}$. We denote this subalgebra by $B_1 \widehat{\otimes}^{(A)} B_2$ or simply $B_1 \widehat{\otimes}^{(\cdot)} B_2$.

Taking into account Lemma 2.6 and Proposition 2.9, we obtain the following result.

Corollary 2.10. *Let A_1 and A_2 be normed algebras, and $A = A_1 \widehat{\otimes}_\gamma A_2$. Then*

(i) *If B_i is a normed subalgebra of A_i for $i = 1, 2$, then $B := B_1 \widehat{\otimes}^{(\cdot)} B_2$ is a Banach subalgebra of A.*
(ii) *If I_i is a flexible ideal of A_i for $i = 1, 2$, then $I := I_1 \widehat{\otimes}^{(\cdot)} I_2$ is a Banach ideal of A and its norm (inherited from the respective quotient) is flexible.*

2.4.3. Quotients of tensor products.

Proposition 2.11. *Let A_1 and A_2 be normed algebras and $A = A_1 \widehat{\otimes} A_2$. Let J_i be ideals of A_i for $i = 1, 2$, and let $J = J_1 \otimes A_2 + A_1 \otimes J_2$. Then*

(i) *The closure of J in A is an ideal of A, and $A/\overline{J}$ is topologically isomorphic to $B = \left(A_1/\overline{J_1}\right) \widehat{\otimes} \left(A_2/\overline{J_2}\right)$.*
(ii) *If I_i are closed ideals of A_i containing J_i, and if $I := I_1 \widehat{\otimes}^{(\cdot)} A_2 + A_1 \widehat{\otimes}^{(\cdot)} I_2$ is a flexible sum of Banach ideals in A, then the closure of J in I is an ideal of I and $I/\overline{J}$ is topologically isomorphic to the algebra*

$$Q = \left(I_1/\overline{J_1}\right) \widehat{\otimes}^{(\cdot)} \left(A_2/\overline{J_2}\right) + \left(A_1/\overline{J_1}\right) \widehat{\otimes}^{(\cdot)} \left(I_2/\overline{J_2}\right)$$

taken with the norm of the flexible sum of Banach ideals in $(A_1/\overline{J_1}) \widehat{\otimes} (A_2/\overline{J_2})$.

Proof. (i) The first statement follows from the fact that J is an ideal of the algebraic tensor product $A_1 \otimes A_2$. To show the second one, assume first that A_1 and A_2 are Banach algebras, and that J_i is a closed ideal of A_i for $i = 1, 2$.

As usual, we denote by q_{J_i} the standard epimorphisms from A_i to A_i/J_i and by q_J the standard epimorphism from $A = A_1 \widehat{\otimes} A_2$ to $A/\overline{J}$. Setting

$$\phi((a_1 + J_1) \otimes (a_2 + J_2)) = q_J(a_1 \otimes a_2),$$

we obtain a bounded homomorphism $\phi : (A_1/J_1) \widehat{\otimes} (A_2/J_2) \to A/\overline{J}$ such that the diagram

$$\begin{array}{ccc} & A/\overline{J} & \\ {\scriptstyle q_J} \nearrow & & \nwarrow {\scriptstyle \phi} \\ A & \xrightarrow{\; q_{J_1} \otimes q_{J_2} \;} & (A_1/J_1) \widehat{\otimes} (A_2/J_2) \end{array}$$

is commutative.

Since q_J is surjective, ϕ is surjective. On the other hand, it is easy to see that $q_{J_1} \otimes q_{J_2}$ is surjective. So, if $\phi(z) = 0$ for some $z \in (A_1/J_1) \widehat{\otimes} (A_2/J_2)$ then $z = (q_{J_1} \otimes q_{J_2})(a)$ for some $a \in A$. In fact, we have that $a \in \overline{J}$ by the commutativity of the diagram. But $J \subset \ker(q_{J_1} \otimes q_{J_2})$, whence $z = 0$. This implies that ϕ is injective. Thus ϕ establishes a bounded isomorphism of B and $A/\overline{J}$. By the Banach Theorem, this isomorphism is topological.

In the general case, passing to completions of A_i and to closures of J_i and applying Proposition 2.6 and simple identifications in completions of quotients of normed algebras, we get the result.

(ii) Follows from a similar analysis of the commutative diagram

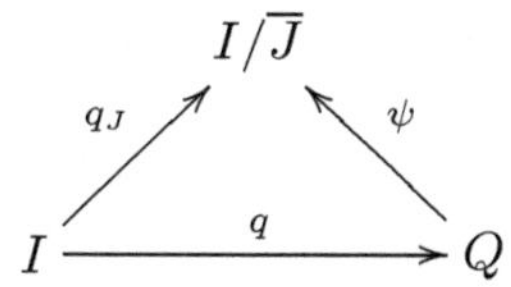

where q is the map sending $p_1 \otimes a_2 + a_1 \otimes p_2$ to $(p_1/\overline{J_1}) \otimes (a_2/\overline{J_2}) + (a_1/\overline{J_1}) \otimes (p_2/\overline{J_2})$ for all $p_i \in I_i$, $a_i \in A_i$ $(i = 1, 2)$. The existence of ψ is evident, surjectivity of q can be verified in a standard way. □

3. Tensor Jacobson radical

3.1. Tensor spectral radius of a summable family

Let A be a normed algebra. We will call by *families* arbitrary sequences of elements of A; two families are *equivalent* (write $\{a_n\}_1^\infty \simeq \{b_n\}_1^\infty$) if one of them can be obtained from the other by renumbering. The equivalence classes can be considered as countable *generalized subsets* [29]: to characterize the class determined by a sequence one have only to indicate which elements of A come into the sequence and how many times.

By definition [29, Section 3.4], a *generalized subset* S of A is a cardinal-valued function $\varkappa_S$ defined on A. The set $\{a \in A : \varkappa_S(a) > 0\}$ is called a *support* of S. One can regard usual subsets $N \subset A$ as generalized ones, identifying the indicator $\varkappa_N$ of N with N. A generalized subset S of A is *countable* if its support and $\varkappa_S(a)$ are (finite or) countable for every a from the support of S.

Let S and P be generalized subsets of A. The *inclusion* $S \subset P$ means

$$\varkappa_S(a) \leq \varkappa_P(a)$$

for every $a \in A$.

We define the *disjoint union* $S \sqcup P$ of generalized subsets of A by

$$\varkappa_{S \sqcup P}(a) = \varkappa_S(a) + \varkappa_P(a)$$

for every $a \in A$. Disjoint union of a collection of generalized subsets is defined similarly. In particular, for an integer $n > 0$, the disjoint union of n copies of S will be denoted by $n \bullet S$.

We define the product SP of generalized subsets of A by

$$\varkappa_{SP}(a) = \sum_{(b,c) \in A \times A,\, bc = a} \varkappa_S(b) \varkappa_P(c)$$

for every $a \in A$.

Given a generalized subset S of A, put

$$\eta(S) = \sum_{a\in A} \varkappa_S(a)\|a\|$$

and

$$\|S\| = \sup_{\varkappa_S(a)>0} \{\|a\| : a \in A,\ \varkappa_S(a) > 0\}.$$

If $\eta(S) < \infty$ then S is called *summable*, and if $\|S\| < \infty$ then S is called *bounded.*

To each sequence $M = \{a_n\}_{n=1}^{\infty}$ in A there corresponds a countable generalized subset $S = S_{(M)}$ by the rule

$$\varkappa_S(a) = \operatorname{card}\{n : a_n = a\}$$

for every $a \in A$. We say that M is a *representative* of S. In terms of representatives $M = \{a_n\}_1^{\infty}$ and $N = \{b_n\}_1^{\infty}$ the family MN corresponds to the two-index sequence $\{a_n b_m\}_{n,m=1}^{\infty}$ which can be renumbered in an arbitrary way, while $M \sqcup N$ corresponds to the sequence $\{c_n\}_1^{\infty}$ with $c_{2k-1} = a_k$, $c_{2k} = b_k$. It is obvious in this context that

$$MN \simeq \sqcup_{i=1}^{\infty} a_i N \simeq \sqcup_{j=1}^{\infty} M b_j,$$

where $aN = \{ab_n\}_1^{\infty}$ and $Mb = \{a_n b\}_1^{\infty}$ as usual. In particular,

$$M(N_1 \sqcup N_2) \simeq MN_1 \sqcup MN_2 \text{ and } (M_1 \sqcup M_2)N \simeq M_1 N \sqcup M_2 N$$

for any families M_i and N_i, $i = 1, 2$, and

$$(MN)K \simeq M(NK) \tag{3.1}$$

for any families in A. Set $M^1 \simeq M$, $M^n \simeq M^{n-1}M$ for every $n > 0$. By (3.1)

$$M^{n+m} \simeq M^n M^m \tag{3.2}$$

for every $n, m \in \mathbb{N}$.

For two families M and N in A, we say that M is a *subfamily* of N (write $M \sqsubset N$) if $S_{(M)} \subset S_{(N)}$ for corresponding generalized subsets of A.

Now let S be a summable generalized subset of A. This is equivalent to the condition that S has a representative M in $\ell_1(A)$, i.e., $S = S_{(M)}$ for some $M \in \ell_1(A)$. Using this and setting $\eta(M) = \eta\left(S_{(M)}\right)$, we simply write "a family $M = \{a_n\}_1^{\infty}$ in A is summable". Moreover,

$$\|M\|_{\ell_1(A)} = \eta(M) = \eta(N)$$

for every $N \simeq M$.

Let M and N be summable families in A. It is evident that

$$\eta(M \sqcup N) = \eta(M) + \eta(N)$$

and

$$\eta(MN) \leqslant \eta(M)\eta(N). \tag{3.3}$$

We obtain by (3.2) and (3.3) that

$$\eta(M^{n+m}) \leqslant \eta(M^n)\eta(M^m) \tag{3.4}$$

for every $n, m \in \mathbb{N}$. It follows from (3.4) that, for every summable family M, there exists a limit

$$\rho_t(M) = \lim(\eta\,(M^n))^{1/n} = \inf(\eta(M^n))^{1/n}.$$

The number $\rho_t(M)$ is called a *tensor spectral radius* of M.

As $(M^m)^n \simeq M^{mn}$ for every $n, m \in \mathbb{N}$, then

$$\rho_t(M^m)^{1/m} = (\lim_n(\eta((M^m)^n))^{1/n})^{1/m} = \lim_n(\eta(M^{mn}))^{1/nm} = \rho_t(M). \qquad (3.5)$$

Now let B be a normed algebra, and let S be a bounded countable generalized subset of B. Then S has a representative L in $\ell_\infty(B)$, i.e., $S = S_{(L)}$ for some $L \in \ell_\infty(B)$. Setting $\|L\| = \|S_{(L)}\|$, we write "a family $L = \{b_n\}_1^\infty$ in B is bounded". Moreover,

$$\|L\|_{\ell_\infty(B)} = \|L\| = \|K\|$$

for every $K \simeq L$. A usual countable subset N of B is bounded if and only if $\sup_{b\in N}\|b\| < \infty$.

Let L and K be bounded families in K. It is evident that

$$\|L \sqcup K\| = \max\{\|L\|, \|K\|\} \text{ and } \|LK\| \le \|L\|\,\|K\|.$$

It follows as above that, for every bounded family L, there is a limit

$$\rho(L) = \lim(\|L^n\|)^{1/n} = \inf(\|L^n\|)^{1/n}.$$

The number $\rho(L)$ is called a *joint spectral radius* of L. It is clear that $\rho(L^m)^{1/m} = \rho(L)$ for every $m \in \mathbb{N}$.

Let A and B be normed algebras, $M = \{a_n\}_1^\infty \in \ell_1(A)$ and $L = \{b_n\}_1^\infty \in \ell_\infty(B)$. Let $M_\otimes L$ denote an element of $A\widehat{\otimes}B$ which is equal to $\sum_{n=1}^\infty a_n \otimes b_n$. It is clear (see also Section 2.4.1) that for every element $z \in A\widehat{\otimes}B$ there are $M \in \ell_1(A)$ and $L \in \ell_\infty(B)$ such that $z = M_\otimes L$.

The following theorem justifies the term "tensor spectral radius".

Theorem 3.1. *Let A be a normed algebra. Then*

(i) *$\rho\,(M_\otimes L) \le \rho_t(M)\rho(L)$ for every normed algebra B, $M \in \ell_1(A)$ and $L \in \ell_\infty(B)$.*

(ii) *There are a unital Banach algebra B, $L \in \ell_\infty(B)$, and a bounded linear operator $T : M \longmapsto M_\otimes L$ from $\ell_1(A)$ into $A\widehat{\otimes}B$ such that $\|M_\otimes L\| = \eta(M)$ and $\rho(M_\otimes L) = \rho_t(M)$ for every $M \in \ell_1(A)$.*

Proof. (i) Let $L = \{b_n\}_1^\infty \in \ell_\infty(B)$. Then for every $M = \{a_n\}_1^\infty \in \ell_1(A)$ we have that

$$\left\|(M_\otimes L)^k\right\| = \gamma\left((M_\otimes L)^k\right) \le \sum_{n_1,\dots,n_k} \|a_{n_1}\cdots a_{n_k}\|\,\|b_{n_1}\cdots b_{n_k}\| \le \eta(M^k)\left\|L^k\right\|$$

Taking k-roots and passing to limits, we obtain that $\rho\,(M_\otimes L) \le \rho_t(M)\rho(L)$.

(ii) Let G be the free unital semigroup with a countable set $W = \{w_k\}_{k\geqslant 1}$ of generators. That is $G = \cup_{m\geqslant 0}W_m$, where $W_0 = 1$, W_m is the direct product of m copies of W realized as the set of 'words' $w_{k_1}w_{k_2}\dots w_{k_m}$ of the length m, and the multiplication is lexical. Let $B = \ell_1(G)$ be the corresponding semigroup algebra.

Let $L = \{w_n\}_1^\infty$. For any $M = \{a_n\}_1^\infty \in \ell_1(A)$, we have that

$$M_\otimes L = \sum_{k=1}^{\infty} a_k \otimes w_k.$$

Then $T : M \longmapsto M_\otimes L$ is a bounded linear operator from $\ell_1(A)$ into $A\widehat{\otimes}B$ and

$$T(M)^n = \sum_{k_1,\dots,k_n} a_{k_1} \dots a_{k_n} \otimes w_{k_1} \dots w_{k_n}.$$

Since $A\widehat{\otimes}\ell_1(G)$ is isometrically isomorphic via the map defined by $(a\otimes f)(g) \longmapsto f(g)a$ to the Banach algebra $\ell_1(G, A)$ of all summable A-valued functions on G, then

$$\|T(M)^n\| = \sum_{k_1,\dots,k_n} \|a_{k_1} \dots a_{k_n}\| = \eta(M^n).$$

It follows that

$$\rho(M_\otimes L) = \rho_t(M). \qquad \square$$

We write $\eta_{\|\cdot\|}(M)$ instead of $\eta(M)$ if there is a necessity to indicate which norm in A is meant.

Proposition 3.2. *Let M be a summable family in a normed algebra A. Then $\rho_t(M)$ doesn't change if the norm on A is changed by an equivalent norm.*

Proof. If $\|\cdot\| \le t\,\|\cdot\|'$ for some $t > 0$, then $\lim \eta_{\|\cdot\|}(M^m)^{1/m} \le \lim \eta_{\|\cdot\|'}(M^m)^{1/m}$, so that the opposite inequality for norms implies the equality of limits. $\square$

For summable families $M = \{a_n\}$ and $N = \{b_n\}_1^\infty$, let $M * N = \{c_n\}$ denote the convolution of M and N: $c_n = \sum_{i+j=n+1} a_i b_j$ for every $n > 0$.

Proposition 3.3. *If $M = \{a_n\}_1^\infty$ and $N = \{b_n\}_1^\infty$ are summable families in A then $\rho_t(M * N) \le \rho_t(MN) = \rho_t(NM)$ and $\rho_t(M + N) \le \rho_t(M \sqcup N)$.*

Proof. Note that $\eta\left((MN)^{n+1}\right) \le \eta(M)\,\eta((NM)^n)\,\eta(N)$ for every n. This implies that $\rho_t(MN) \le \rho_t(NM)$. Changing M and N by places, we have the equality.

We have that

$$\begin{aligned}
\eta\left((M*N)^k\right) &= \sum_{n_1,\dots,n_k} \left\| \left(\sum_{i_i+j_i=n_1+1} a_{i_1} b_{j_1}\right) \cdots \left(\sum_{i_k+j_k=n_k+1} a_{i_k} b_{j_k}\right) \right\| \\
&\le \sum_{n_1,\dots,n_k} \sum_{i_i+j_i=n_1+1} \cdots \sum_{i_k+j_k=n_k+1} \|a_{i_1} b_{j_1} \cdots a_{i_k} b_{j_k}\| \\
&= \sum_{n_1,\dots,n_{2k}} \|a_{n_1} b_{n_2} \cdots a_{n_{2k-1}} b_{n_{2k}}\| = \eta\left((MN)^k\right)
\end{aligned}$$

for every $k > 0$, whence $\rho_t(M * N) \le \rho_t(MN)$.

Further, $M+N=\{a_n+b_n\}_1^\infty$ and

$$(M+N)^k \simeq \{(a_{n_1}+b_{n_1})\cdots(a_{n_k}+b_{n_k})\}_{n_1,\dots,n_k=1}^\infty = \{a_{n_1}\cdots a_{n_k}+b_{n_1}a_{n_2}\cdots a_{n_k}+\cdots+b_{n_1}\cdots b_{n_k}\}_{n_1,\dots,n_k=1}^\infty .$$

Then

$$\begin{aligned}\eta\left((M+N)^k\right) &= \sum_{n_1,\dots,n_k}\|a_{n_1}\cdots a_{n_k}+b_{n_1}a_{n_2}\cdots a_{n_k}+\cdots+b_{n_1}\cdots b_{n_k}\|\\ &\le \sum_{n_1,\dots,n_k}(\|a_{n_1}\cdots a_{n_k}\|+\|b_{n_1}a_{n_2}\cdots a_{n_k}\|+\cdots+\|b_{n_1}\cdots b_{n_k}\|)\\ &= \sum_{n_1,\dots,n_k}\|a_{n_1}\cdots a_{n_k}\|+\sum_{n_1,\dots,n_k}\|b_{n_1}a_{n_2}\cdots a_{n_k}\|+\cdots\\ &\quad\cdots+\sum_{n_1,\dots,n_k}\|b_{n_1}\cdots b_{n_k}\|\\ &= \eta\left(M^k\right)+\eta\left(NM^{k-1}\right)+\cdots+\eta\left(N^k\right)\\ &= \eta\left(M^k\sqcup NM^{k-1}\sqcup\cdots\sqcup N^k\right)=\eta\left((M\sqcup N)^k\right)\end{aligned}$$

for every $k>0$, whence $\rho_t(M+N)\le\rho_t(M\sqcup N)$. □

As A is embedded into A^1, let M^0 be $\{x_n\}_1^\infty$ with $x_n=0$ for every $n>1$ and $x_1=1$, the identity element of A^1. Note that $M^0M^0\simeq M^0\simeq N^0$, $M^0N\simeq NM^0\simeq N$ for any family N in A, and $\eta\left(M^0\right)=1$.

We say that families M and N in A *commute* if $MN\simeq NM$. This of course doesn't mean that elements of M commute with elements of N. But the reverse statement clearly keeps: if each element of M commutes with each element of N then M and N commute. In particular, if N consists of elements of the center of A then M and N commute.

Proposition 3.4. *Let M and N be summable families in A. If M and N commute then $\rho_t(MN)\le\rho_t(M)\rho_t(N)$ and $\rho_t(M\sqcup N)\le\rho_t(M)+\rho_t(N)$.*

Proof. Indeed, $(NM)^n\simeq N^nM^n$ and $\eta((NM)^n)=\eta(N^nM^n)\le\eta(N^n)\eta(M^n)$ for every n. Taking n-roots and passing to limits, we obtain that

$$\rho_t(MN)\le\rho_t(M)\rho_t(N).$$

It is easy to see that

$$(M\sqcup N)^n\simeq\sqcup_{i=0}^n\left(C_n^i\bullet\left(M^iN^{n-i}\right)\right)$$

for every $n>0$, whence

$$\eta((M\sqcup N)^n)=\sum_{i=0}^n C_n^i\eta\left(M^iN^{n-i}\right)\le\sum_{i=0}^n C_n^i\eta\left(M^i\right)\eta\left(N^{n-i}\right).$$

Let $\varepsilon > 0$, and take $s \geq 1$ such that $\eta\left(M^i\right) \leq s\left(\rho_t\left(M\right)+\varepsilon\right)^i$ and $\eta\left(N^i\right) \leq s\left(\rho_t\left(N\right)+\varepsilon\right)^i$ for every $i \in \mathbb{N}$. Then

$$\begin{aligned}\eta\left(\left(M \sqcup N\right)^n\right) &\leq s^2 \sum_{i=0}^{n} C_n^i \left(\rho_t\left(M\right)+\varepsilon\right)^i \left(\rho_t\left(N\right)+\varepsilon\right)^{n-i} \\ &= s^2\left(\rho_t\left(M\right)+\rho_t\left(N\right)+2\varepsilon\right)^n\end{aligned}$$

for every $n > 0$. Taking n-roots and passing to limits, we obtain that

$$\rho_t\left(M \sqcup N\right) \leq \rho_t\left(M\right)+\rho_t\left(N\right)+2\varepsilon$$

As ε is arbitrary, we have that $\rho_t\left(M \sqcup N\right) \leq \rho_t\left(M\right)+\rho_t\left(N\right)$. □

3.2. Absolutely convex hulls and tensor quasinilpotent families

Let A be a normed algebra. A summable family M of elements of A is called *tensor quasinilpotent* if $\rho_t\left(M\right) = 0$.

The following result is an immediate consequence of Theorem 3.1(i).

Corollary 3.5. *If a family M in A is tensor quasinilpotent then for each bounded family L in a normed algebra B the element $M_{\otimes}L$ is quasinilpotent in $A\widehat{\otimes}B$.*

For a summable family $M = \{a_n\}_1^\infty$, let $\mathrm{abs}_t\left(M\right)$ denote the set of all families $N = \{b_n\}_1^\infty$ such that $b_m = \sum_{n=1}^\infty t_{nm}a_n$, where the sequences $\{t_{nm}\}_{m=1}^\infty$ of complex numbers satisfy the condition $\sum_{m=1}^\infty |t_{nm}| \leq 1$. We call $\mathrm{abs}_t\left(M\right)$ the *absolutely convex hull* of M. To justify the term, note that $\mathrm{abs}_t\left(M\right)$ is a closed absolutely convex subset of $\ell_1\left(A\right)$.

Proposition 3.6. *If $M = \{a_n\}_1^\infty$ is a summable family of elements of a normed algebra A then $\rho_t\left(N\right) \leq \rho_t\left(M\right)$ for any $N = \{b_n\}_1^\infty \in \mathrm{abs}_t\left(M\right)$.*

Proof. Indeed,

$$\begin{aligned}\eta(N^k) &= \sum_{m_1,\dots,m_k} \|b_{m_1}\cdots b_{m_k}\| = \sum_{m_1,\dots,m_k}\left\|\sum_{n_1,\dots,n_k} t_{n_1m_1}\cdots t_{n_km_k}a_{n_1}\cdots a_{n_k}\right\| \\ &\leq \sum_{m_1,\dots,m_k}\sum_{n_1,\dots,n_k} |t_{n_1m_1}\cdots t_{n_km_k}|\,\|a_{n_1}\cdots a_{n_k}\| \\ &= \sum_{n_1,\dots,n_k}\sum_{m_1}|t_{n_1m_1}|\cdots\sum_{m_k}|t_{n_km_k}|\,\|a_{n_1}\cdots a_{n_k}\| \\ &\leq \sum_{n_1,\dots,n_k}\|a_{n_1}\cdots a_{n_k}\| = \eta(M^k)\end{aligned}$$

for every k, whence $\rho_t\left(N\right) \leq \rho_t\left(M\right)$. □

Recall that a set K in a normed space X is called *absolutely convex* if $t_1x_1 + t_2x_2 \in K$ for every integer $n > 0$ and for any $x_1, x_2 \in K$ and $t_1, t_2 \in \mathbb{C}$ with $|t_1|+|t_2| \leq 1$. If K is a compact set in X, then the number $\max\{\|x-y\| : x, y \in K\}$ is called the *diameter* of K and denoted by $\mathrm{diam}\left(K\right)$.

Lemma 3.7. *Let X be a Banach space and let $\{K_n\}$ be a sequence of absolutely convex compact sets in X such that $\sum \operatorname{diam}(K_n) < \infty$. Then $\sum K_n$ is an absolutely convex compact set.*

Proof. It is clear that $\sum K_n$ is absolutely convex. To see that it is compact, note that the direct product $K = \Pi_{n=1}^{\infty} K_n$ is compact. Since $0 \in K_n$, we get that $\|a\| \le \operatorname{diam}(K_n)/2$ for $a \in K_n$. So there are numbers $\alpha_n > 0$ with $\|a\| \le \alpha_n$ for $a \in K_n$, such that $\sum \alpha_n < \infty$. It follows that the map $\varphi : K \longrightarrow X$ defined by the formula

$$\varphi(\{a_n\}_{n=1}^{\infty}) = \sum_{n=1}^{\infty} a_n,$$

is continuous. As $\sum K_n = \varphi(K)$, it is compact. □

Let $M = \{a_n\}_1^{\infty}$ be a summable family in a Banach algebra A, and let $\Omega(M)$ denote the set of all elements of the form $\sum t_n a_n$ for complex numbers $|t_n| \le 1$.

Corollary 3.8. *If M is a summable family in a Banach algebra A then $\Omega(M)$ is an absolutely convex compact set such that $\rho(a) \le \rho_t(M)$ for any $a \in \Omega(M)$. Moreover, $\Omega(M) = \{\sum b_n : \{b_n\}_1^{\infty} \in \operatorname{abs}_t(M)\}$.*

Proof. Indeed, $\Omega(M)$ is the countable sum of absolutely convex compact sets $\{ta_n : |t| \le 1\}$, the sum of whose diameters is finite. So $\Omega(M)$ is a convex compact set in A. Further, for any $a = \sum t_n a_n \in \Omega(M)$, we have that

$$\|a^k\| \le \sum_{n_1,\dots,n_k} \|t_{n_1} \cdots t_{n_k} a_{n_1} \cdots a_{n_k}\| \le \sum_{n_1,\dots,n_k} \|a_{n_1} \cdots a_{n_k}\| = \eta(M^k)$$

for every $k \in \mathbb{N}$. So $\rho(a) \le \rho_t(M)$.

The last assertion easily follows from well-known properties of absolutely summable series. □

Corollary 3.9. *Let $M = \{a_n\}_1^{\infty}$ be a tensor quasinilpotent family in a Banach algebra A. Then every element of $\Omega(M)$ is quasinilpotent.*

Lemma 3.10. *If $M = \{a_n\}_1^{\infty}$ is a summable family in A such that $\rho_t(M) < 1$ then the family $\sqcup_{m=1}^{\infty} M^m$ is summable and $\rho_t(\sqcup_{m=1}^{\infty} M^m) = \rho_t(M)(1 - \rho_t(M))^{-1}$.*

Proof. Let $M_{(k)} \simeq \sqcup_{n=k}^{\infty} M^n$ for any k. Take t satisfying $\rho_t(M) < t < 1$. Then there is an integer p such that $\eta(M^n) \le t^n$ for every $n \ge p$. Then

$$\sum_{n=1}^{\infty} \eta(M^n) < \sum_{n=1}^{p-1} \eta(M^n) + \sum_{n=p}^{\infty} t^n = \sum_{n=1}^{p-1} \eta(M^n) + \frac{t^p}{1-t} < \infty.$$

As $\eta(\sqcup_{n=1}^{\infty} M^n) = \sum_{n=1}^{\infty} \eta(M^n) < \infty$, $M_{(1)}$ is summable.

Since $M_{(2)} \simeq M M_{(1)} \simeq M_{(1)} M$ and $M_{(1)} \simeq M \sqcup M_{(2)} \simeq M\left(M^0 \sqcup M_{(1)}\right)$, we obtain that

$$\rho_t\left(M_{(1)}\right) \le \rho_t(M)\, \rho_t\left(M^0 \sqcup M_{(1)}\right) \le \rho_t(M)\left(\rho_t\left(M_{(1)}\right) + 1\right)$$

by Proposition 3.4, whence $\rho_t\left(M_{(1)}\right) \le \rho_t(M)(1 - \rho_t(M))^{-1}$.

On the other hand, we have, using (3.5), that

$$\begin{aligned}\rho_t(M)^n(1-\rho_t(M))^{-n} &= \left(\rho_t(M)+\rho_t(M)^2+\rho_t(M)^3+\cdots\right)^n\\ &= \rho_t(M)^n + n\rho_t(M)^{n+1}+\frac{n(n+1)}{2}\rho_t(M)^{n+2}+\cdots\\ &\overset{(3.5)}{=}\rho_t(M^n)+n\rho_t\left(M^{n+1}\right)+\frac{n(n+1)}{2}\rho_t\left(M^{n+2}\right)+\cdots\\ &\le \eta(M^n)+n\eta\left(M^{n+1}\right)+\frac{n(n+1)}{2}\eta\left(M^{n+2}\right)+\cdots\\ &= \eta\left(M^n\sqcup n\bullet M^{n+1}\sqcup\frac{n(n+1)}{2}\bullet M^{n+2}\sqcup\cdots\right)\\ &= \eta\left(\left(M\sqcup M^2\sqcup M^3\sqcup\cdots\right)^n\right)=\eta\left(M_{(1)}^n\right)\end{aligned}$$

for every $n>0$, whence $\rho_t(M)(1-\rho_t(M))^{-1}\le\rho_t\left(M_{(1)}\right)$. □

Corollary 3.11. *If $M=\{a_n\}_1^\infty$ is a tensor quasinilpotent family in a normed algebra A then the subalgebra generated by M consists of quasinilpotents.*

Proof. Indeed, $\rho_t\left(\sqcup_{m=1}^\infty M^m\right)=0$ by Lemma 3.10, and every element of the subalgebra lies in $\cup_{t>0}t\,\Omega\left(\sqcup_{m=1}^\infty M^m\right)$ which consists of quasinilpotents by Corollary 3.9. □

3.3. Upper semicontinuity and subharmonicity of the tensor spectral radius

Let A be a normed algebra. Since sequences in $\ell_1(A)$ determine summable families, the tensor spectral radius can be considered as a function on $\ell_1(A)$. We are going to show that this function is upper semicontinuous and subharmonic.

If G is a subset of A, let $\mathrm{F}_1(G)$ be the set of all summable families $M=\{a_n\}_1^\infty$ with all $a_n\in G$. In particular, $\mathrm{F}_1(A)=\ell_1(A)$. Clearly $\mathrm{F}_1(G)$ is a metric space with respect to the metric $\mathrm{d}(M,N)=\eta(M-N)$ induced by the norm on $\ell_1(A)$. If A is a Banach algebra and G is a closed subset of A then $\mathrm{F}_1(G)$ is a complete metric space.

Now we are able to establish the upper semicontinuity of the tensor spectral radius.

Proposition 3.12. *Let $M\in\ell_1(A)$. For every $\varepsilon>0$, there is $\delta>0$ such that $\rho_t(N)\le\rho_t(M)+\varepsilon$ for every $N\in\ell_1(A)$ satisfying $\mathrm{d}(N,M)<\delta$.*

Proof. Let $T:M\longmapsto M_\otimes L$ be the map defined in Theorem 3.1(ii). As the usual spectral radius is upper semicontinuous, to any $\varepsilon>0$, there corresponds $\delta>0$ such that $\rho(T(N))\le\rho(T(M))+\varepsilon$ for every $N\in\ell_1(A)$ satisfying $\|T(N)-T(M)\|<\delta$. By Theorem 3.1(ii), we get that $\rho(T(N))=\rho_t(N)$, $\rho(T(M))=\rho_t(M)$ and $\|T(N)-T(M)\|=\|T(N-M)\|=\eta(M-N)=\mathrm{d}(M,N)$. □

Let F be an analytic function on an open set $\mathcal{D} \subset \mathbb{C}$ with values in $\ell_1(A)$. This means that for every $\lambda \in \mathcal{D}$ there is $F'(\lambda) \in \ell_1(A)$ such that

$$F'(\lambda) = \lim_{\mu \to \lambda} \frac{F(\mu) - F(\lambda)}{\mu - \lambda}$$

in the norm of $\ell_1(A)$. Clearly F as an $\ell_1(A)$-valued function induces the family $\{f_n\}_1^\infty$ of A-valued functions on $\mathcal{D}$: $F = \{f_n\}_1^\infty$. These functions are analytic since $\lambda \longmapsto F(\lambda) = \{f_n(\lambda)\}_1^\infty$ is analytic on $\mathcal{D}$. So one can write $F' = \{f'_n\}_1^\infty$, etc.

Let now $F = \{f_n\}_1^\infty$ and $\Psi = \{\psi_n\}_1^\infty$ be $\ell_1(A)$-valued functions on $\mathcal{D}$. Let $F\Psi$ denote some sequence $\{\varphi_k\}_1^\infty$ of functions on $\mathcal{D}$ which is obtained from $\{f_n\psi_m\}_{n,m=1}^\infty$ by renumbering. We will assume that the renumbering for this operation is fixed, so $F\Psi$ is defined correctly. In such a case the functions φ_k are A-valued functions on $\mathcal{D}$, and we write that $F\Psi = \{\varphi_k\}$ is an $\ell_1(A)$-valued function on $\mathcal{D}$.

Lemma 3.13. *If $F = \{f_n\}_1^\infty$ and $\Psi = \{\psi_n\}_1^\infty$ are analytic $\ell_1(A)$-valued function on $\mathcal{D}$ then $F\Psi$ is an analytic $\ell_1(A)$-valued function on $\mathcal{D}$.*

Proof. Indeed, the derivation $(F\Psi)'$ exists and is clearly obtained from two-index sequence $\{f'_n\psi_m + f_n\psi'_m\}_{n,m=1}^\infty$ by the same renumbering as $F\Psi$ from $\{f_n\psi_m\}_{n,m=1}^\infty$. Moreover,

$$\begin{aligned}
\eta\left((F\Psi)'(\lambda)\right) &= \sum_{n,m} \|f'_n(\lambda)\psi_m(\lambda) + f_n(\lambda)\psi'_m(\lambda)\| \\
&\leq \sum_{n,m} (\|f'_n(\lambda)\| \|\psi_m(\lambda)\| + \|f_n(\lambda)\| \|\psi'_m(\lambda)\|) \\
&= \sum_n \|f'_n(\lambda)\| \sum_m \|\psi_m(\lambda)\| + \sum_n \|f_n(\lambda)\| \sum_m \|\psi'_m(\lambda)\| \\
&= \eta(F'(\lambda))\,\eta(\Psi(\lambda)) + \eta(F(\lambda))\,\eta(\Psi'(\lambda)) < \infty. \qquad \square
\end{aligned}$$

For an $\ell_1(A)$-valued function $F = \{f_n\}_1^\infty$, let $F^1 = F$ and $F^m = F^{m-1}F$ for $m > 1$, so that $F^m = \{\phi_n\}_1^\infty$ is an $\ell_1(A)$-valued function for some A-valued functions ϕ_n on $\mathcal{D}$. It follows by induction from the definition of product of two functions that

$$F^m(\lambda) \simeq F(\lambda)^m \tag{3.6}$$

for every $\lambda \in \mathcal{D}$.

Corollary 3.14. *If F is an analytic $\ell_1(A)$-valued function on $\mathcal{D}$ then F^m is an analytic $\ell_1(A)$-valued function on $\mathcal{D}$ for every $m > 0$.*

Proof. Follows from Lemma 3.13 by induction. $\square$

For an $\ell_1(A)$-valued function F, the function $\lambda \longmapsto \eta(F(\lambda)^m)$ doesn't depend on a renumbering. Moreover, it follows from (3.6) that

$$\eta(F(\lambda)^m) = \eta(F^m(\lambda)) \tag{3.7}$$

for every $\lambda \in \mathcal{D}$.

Lemma 3.15. *If F is an analytic $\ell_1(A)$-valued function on $\mathcal{D}$ then the functions $\lambda \longmapsto \eta(F(\lambda)^m)$ and $\lambda \longmapsto \log(\eta(F(\lambda)^m))$ are subharmonic on $\mathcal{D}$ for all m.*

Proof. As $\eta(\cdot)$ is a norm on $\ell_1(A)$, it is well known (see [33]) that $\lambda \longmapsto \eta(F(\lambda))$ is a subharmonic function. Then it follows from Corollary 3.14 and (3.7) that $\lambda \longmapsto \eta(F(\lambda)^m)$ is subharmonic for every m.

Let $F^m = \{\phi_n\}_1^\infty$ and $\beta \in \mathbb{C}$. As

$$|\exp(\beta\lambda)|\,\eta(F(\lambda)^m) = \eta(\{\exp(\beta\lambda)\,\phi_n(\lambda)\}_1^\infty)$$

by (3.7) and $\lambda \longmapsto \{\exp(\beta\lambda)\,\phi_n(\lambda)\}_1^\infty$ determines an analytic $\ell_1(A)$-valued function on $\mathcal{D}$, then, by above, $\lambda \longmapsto |\exp(\beta\lambda)|\,\eta(F(\lambda)^m)$ is subharmonic for every $\beta \in \mathbb{C}$. It follows from Rado's theorem [2, Appendix 2, Theorem 9] that the function $\lambda \longmapsto \log(\eta(F(\lambda)^m))$ is subharmonic on $\mathcal{D}$. □

We use Vesentini's argument for subharmonicity of the usual spectral radius [33] in the following

Theorem 3.16. *If F is an analytic $\ell_1(A)$-valued function on $\mathcal{D}$ then the functions $\log(\rho_t(F)) : \lambda \longmapsto \log(\rho_t(F(\lambda)))$ and $\rho_t(F) : \lambda \longmapsto \rho_t(F(\lambda))$ are subharmonic on $\mathcal{D}$.*

Proof. As $\eta\left(F(\lambda)^{2^{m+1}}\right) \leq \eta\left(F(\lambda)^{2m}\right)^2$, the function $\lambda \longmapsto \log(\rho_t(F(\lambda)))$ is a pointwise limit of the decreasing sequence $\left\{\lambda \longmapsto 2^{-m}\log\left(\eta\left(F(\lambda)^{2m}\right)\right)\right\}$ of subharmonic functions and is therefore a subharmonic function by Theorem 1 of [2, Appendix 2]. Since the function $t \longmapsto \exp(t)$ is convex and positive for $t \in \mathbb{R}$, then the function $\lambda \longmapsto \exp(\log(\rho_t(F(\lambda)))) = \rho_t(F(\lambda))$ is also subharmonic by the same theorem. □

3.4. The ideal $\mathcal{R}_t(A)$ for a normed algebra A

Let a be an element of a normed algebra A, and let $M = \{a_n\}_{n=1}^\infty$ be a summable family in A. Let $\{a\} \sqcup M$ denote the family $\{x_n\}_1^\infty$ with $x_1 = a$ and $x_n = a_{n-1}$ for $n > 1$. Note that the useful relation

$$(\{a\} \sqcup M)(\{b\} \sqcup N) \simeq \{ab\} \sqcup aN \sqcup Mb \sqcup MN \tag{3.8}$$

is valid by our conventions, for any $a, b \in A$ and $M, N \in \ell_1(A)$.

Let $\mathcal{R}_t(A)$ be the set of all $a \in A$ such that $\rho_t(\{a\} \sqcup M) = \rho_t(M)$ for every $M \in \ell_1(A)$. It is evident that $\mathcal{R}_t(A)$ consists of quasinilpotent elements of A.

Lemma 3.17. *Let $a \in A$. If there is $s > 0$ such that $\rho_t(\{a\} \sqcup M) \leq s\rho_t(M)$ for every $M \in \ell_1(A)$ then $a \in \mathcal{R}_t(A)$.*

Proof. As the function $\mu \longmapsto \rho_t(\{\mu a\} \sqcup M)$ is subharmonic by Theorem 3.16 and bounded on $\mathbb{C}$, it is constant, whence $\rho_t(\{a\} \sqcup M) = \rho_t(M)$. □

Lemma 3.18. $\mathcal{R}_t(A) = A \cap \mathcal{R}_t(A^1)$. *If A is complete then* $\mathcal{R}_t(A) = \mathcal{R}_t(A^1)$.

Proof. Let A be non-unital and $a \in \mathcal{R}_t(A)$. As $A^1 = A \oplus \mathbb{C}$, for every summable family M in A^1 there are $N \in \ell_1(A)$ and $K \in \ell_1(\mathbb{C})$ such that $M = N + K$. Since N and K commute, then

$$\rho_t(M) = \rho_t(N + K) \leq \rho_t(N \sqcup K) \leq \rho_t(N) + \rho_t(K)$$

by Propositions 3.3 and 3.4. Hence, as $\{\mu a\} \sqcup N$ and $\{0\} \sqcup K$ commute,

$$\begin{aligned}\rho_t(\{\mu a\} \sqcup M) &\leq \rho_t(\{\mu a\} \sqcup N + \{0\} \sqcup K) \\ &\leq \rho_t(\{\mu a\} \sqcup N) + \rho_t(\{0\} \sqcup K) = \rho_t(N) + \rho_t(K) < \infty\end{aligned}$$

for every $\mu \in \mathbb{C}$. Therefore $\mu \longmapsto \rho_t(\{\mu a\} \sqcup M)$ is constant, and as a consequence,

$$\rho_t(\{a\} \sqcup M) = \rho_t(M).$$

As M is arbitrary, $a \in \mathcal{R}_t(A^1)$. So $\mathcal{R}_t(A) \subset \mathcal{R}_t(A^1)$.

On the other hand, $A \cap \mathcal{R}_t(A^1) \subset \mathcal{R}_t(A)$ by definition. So we obtain that

$$\mathcal{R}_t(A) = A \cap \mathcal{R}_t(A^1).$$

Let now A be complete. We show that $\mathcal{R}_t(A^1) \subset A$. Indeed, if $a - \lambda \in \mathcal{R}_t(A^1)$ with $a \in A$ and $\lambda \in \mathbb{C}$ then $a - \lambda$ is a quasinilpotent element of A^1. This means that the spectrum $\sigma(a)$ of a is equal to $\{\lambda\}$. But, as $a \in A$ and A is not unital, $\sigma(a)$ contains zero. Therefore $\lambda = 0$.

So, if A is complete, $\mathcal{R}_t(A^1) \subset A$, whence $\mathcal{R}_t(A^1) = \mathcal{R}_t(A)$ by the above. □

Theorem 3.19. *$\mathcal{R}_t(A)$ is a closed ideal of A.*

Proof. Consider first the case when A has the identity element 1. Let $a, b \in \mathcal{R}_t(A)$. As

$$\rho_t(\{\mu a\} \sqcup M) = |\mu| \, \rho_t(\{a\} \sqcup \mu^{-1} M) = |\mu| \, \rho_t(\mu^{-1} M) = \rho_t(M)$$

for every $M \in \ell_1(A)$ and non-zero $\mu \in \mathbb{C}$, then $\mu a \in \mathcal{R}_t(A)$ for every $\mu \in \mathbb{C}$.

Since $\{2^{-1}(a + b)\} \sqcup M \in \mathrm{abs}_t(\{a\} \sqcup \{b\} \sqcup M)$ for every $M \in \ell_1(A)$, then, by Proposition 3.6,

$$\rho_t(\{2^{-1}(a + b)\} \sqcup M) \leq \rho_t(\{a\} \sqcup \{b\} \sqcup M) = \rho_t(M),$$

whence $a + b \in \mathcal{R}_t(A)$ by Lemma 3.17. So $\mathcal{R}_t(A)$ is a subspace of A.

Let $x \in A$. Then

$$\rho_t(\{\mu a\} \sqcup \{1\} \sqcup \{x\} \sqcup M) \leq t_0 := \rho_t(\{1\} \sqcup \{x\} \sqcup M) < \infty$$

for every $\mu \in \mathbb{C}$. Therefore

$$\rho_t\left((\{\mu a\} \sqcup \{1\} \sqcup \{x\} \sqcup M)^2\right) \leq t_0^2$$

by (3.5). Since $\{\mu a x\} \sqcup M$ is a subfamily of $(\{\mu a\} \sqcup \{1\} \sqcup \{x\} \sqcup M)^2$ in virtue of (3.8), we obtain that

$$\rho_t(\{\mu a x\} \sqcup M) \leq t_0^2$$

for every $\mu \in \mathbb{C}$ and for every $M \in \ell_1(A)$ with t_0 depending only on x and M. Therefore $\mu \longmapsto \rho_t(\{\mu a x\} \sqcup M)$ is bounded on $\mathbb{C}$. As this function is subharmonic, it is constant, whence $ax \in \mathcal{R}_t(A)$, and, similarly, $xa \in \mathcal{R}_t(A)$. Thus $\mathcal{R}_t(A)$ is an ideal of A.

Note that

$$\begin{aligned}\rho_t(\{a+x\} \sqcup M) &\le \rho_t(\{2a\} \sqcup \{2x\} \sqcup M) = \rho_t(\{2x\} \sqcup M)\\ &\le \eta(\{2x\} \sqcup M) = \|2x\| + \eta(M)\end{aligned}$$

for every $a \in \mathcal{R}_t(A)$ and $x \in A$. Now if c is in the closure of $\mathcal{R}_t(A)$, then for every $\mu \in \mathbb{C}$ there are $a \in \mathcal{R}_t(A)$ and $x \in A$ with $\|x\| \le 1$ such that $\mu c = a + x$. Hence

$$\rho_t(\{\mu c\} \sqcup M) = \rho_t(\{a+x\} \sqcup M) \le 2 + \eta(M)$$

for every $M \in \ell_1(A)$. So $\mu \longmapsto \rho_t(\{\mu c\} \sqcup M)$ is bounded and therefore constant, whence $c \in \mathcal{R}_t(A)$. Thus $\mathcal{R}_t(A)$ is a closed ideal of A.

Now assume that A is not unital. We already have proved that $\mathcal{R}_t(A^1)$ is a closed ideal of A^1. Then Lemma 3.18 shows that $\mathcal{R}_t(A)$ is a closed ideal of A. □

Corollary 3.20. *If A is a Banach algebra then $\mathcal{R}_t(A) \subset \mathrm{Rad}(A)$.*

Proof. Indeed, $\mathcal{R}_t(A)$ is an ideal of A consisting of quasinilpotents. So we have that $\mathcal{R}_t(A) \subset \mathrm{Rad}(A)$. □

Theorem 3.21. *If $\rho_t(aM) = 0$ for every $M \in \ell_1(A)$ then $a \in \mathcal{R}_t(A)$.*

Proof. Let first A have the identity element 1. Let $M \in \ell_1(A)$. Multiplying M by a scalar, one can assume that

$$\eta(M) < 1. \tag{3.9}$$

Then $N := \sqcup_{i=0}^{\infty} M^i$ is a summable family in A by Lemma 3.10, where $M^0 = \{x_n\}_{n=1}^{\infty}$ with $x_1 = 1$ and $x_i = 0$ for every $i > 1$ as usual. By condition, we have that $\rho_t(aN) = 0$. Let $\mu \in \mathbb{C}$ be non-zero and take $\varepsilon > 0$ such that $\varepsilon |\mu| < 2^{-1}$. Then there is $t > 0$ such that

$$\eta((aN)^n) \le t\varepsilon^n \tag{3.10}$$

for every $n > 0$. We have that

$$\begin{aligned}\eta((\{\mu a\} \sqcup M)^n) &= \sum_{i=0}^{n} |\mu|^i \sum_{\sum_{k=0}^{i} m_k = n-i} \eta(M^{m_0} a M^{m_1} \cdots a M^{m_i})\\ &\overset{(3.9)}{\le} \eta(M^n) + \sum_{i=1}^{n} |\mu|^i \sum_{\sum_{k=0}^{i} m_k = n-i} \eta(a M^{m_1} \cdots a M^{m_i})\end{aligned}$$

for every $n > 0$. As, for every $i > 0$, the number of summands $\eta(aM^{m_1} \cdots aM^{m_i})$ is less than or equal to 2^n and every such summand is less than or equal to

$\eta\left((aN)^i\right)$, then we obtain that

$$\eta\left((\{\mu a\}\sqcup M)^n\right)\leq\eta\left(M^n\right)+2^n\sum_{i=1}^{n}|\mu|^i\,\eta\left((aN)^i\right)\overset{(3.10)}{\leq}\eta\left(M^n\right)+2^n t_0\sum_{i=1}^{n}|\mu|^i\,\varepsilon^i$$
$$\leq\eta\left(M^n\right)+2^n t\leq 2\max\left\{\eta\left(M^n\right),2^n t\right\}$$

Taking n-roots and passing to limits, we get $\rho_t\left(\{\mu a\}\sqcup M\right)\leq\max\left\{\rho_t\left(M\right),2\right\}$ for every $\mu\in\mathbb{C}$. As the function $\mu\longmapsto\rho_t\left(\{\mu a\}\sqcup M\right)$ is bounded and subharmonic, it is constant. Therefore

$$\rho_t\left(\{a\}\sqcup M\right)=\rho_t\left(M\right).$$

As M is arbitrary, $a\in\mathcal{R}_t\left(A\right)$.

Now assume that A is not unital. Then, for each $M\in\ell_1\left(A^1\right)$, the family $K=MaM$ belongs to $\ell_1\left(A\right)$, and $\rho_t((aM)^2)=\rho_t(aK)=0$ by condition. By (3.5), we obtain that $\rho_t(aM)=0$ for every $M\in\ell_1\left(A^1\right)$, whence $a\in R_t\left(A^1\right)$ by the proof above. Now the result follows from Lemma 3.18. □

3.5. Tensor quasinilpotent algebras and ideals

Let A be a normed algebra. A subset G of A is called a *tensor quasinilpotent set* if all summable families with elements in G are tensor quasinilpotent. A *tensor quasinilpotent ideal* in A is an ideal which is a tensor quasinilpotent subset of A.

Theorem 3.22. *$\rho_t\left(M\sqcup N\right)=\rho_t\left(N\right)$ for every $M\in\ell_1\left(\mathcal{R}_t\left(A\right)\right)$ and $N\in\ell_1\left(A\right)$.*

Proof. Let $N\in\ell_1\left(A\right)$, $M=\{a_n\}_{n=1}^{\infty}\in\ell_1\left(\mathcal{R}_t\left(A\right)\right)$ and $M_k=\{a_n\}_{n=k}^{\infty}$ for every integer $k>0$. For every $\varepsilon>0$ and $\mu\in\mathbb{C}$, there is $k>0$ such that $\eta\left(\mu M_k\right)<\varepsilon$. Then $\mu\longmapsto\mu M\sqcup N$ is an analytic function and

$$\rho_t\left(\mu M\sqcup N\right)=\rho_t\left(\mu M_2\sqcup N\right)=\cdots=\rho_t\left(\mu M_k\sqcup N\right)\leq\eta\left(\mu M_k\sqcup N\right)$$
$$=\eta\left(\mu M_k\right)+\eta\left(N\right)<\eta\left(N\right)+\varepsilon.$$

So $\mu\longmapsto\rho_t\left(\mu M\sqcup N\right)$ is bounded and therefore constant. Hence we obtain that $\rho_t\left(M\sqcup N\right)=\rho_t\left(N\right)$. □

As a consequence, we obtain the following

Corollary 3.23. *$\mathcal{R}_t\left(A\right)$ is a tensor quasinilpotent ideal.*

Corollary 3.24. *$\rho_t\left(M+N\right)=\rho_t\left(N\right)$ and $\rho_t\left(M*N\right)=\rho_t\left(MN\right)=0$ for every $M\in\ell_1\left(\mathcal{R}_t\left(A\right)\right)$ and $N\in\ell_1\left(A\right)$.*

Proof. Let $\mu\in\mathbb{C}$. Then $\rho_t\left(\mu M+N\right)\leq\rho_t\left(\mu M\sqcup N\right)=\rho_t\left(N\right)$ by Proposition 3.3 and Theorem 3.22. As $\mu\longmapsto\rho_t\left(\mu M+N\right)$ is subharmonic and bounded on $\mathbb{C}$, it is constant, whence $\rho_t\left(M+N\right)=\rho_t\left(N\right)$.

Since $MN\in\ell_1\left(\mathcal{R}_t\left(A\right)\right)$, then $\rho_t\left(MN\right)=0$ by Corollary 3.23. Then we obtain that $\rho_t\left(M*N\right)=0$ by Proposition 3.3. □

Corollary 3.25. *Let A be a normed algebra and $a\in A$. The following conditions are equivalent:*

(i) $a \in \mathcal{R}_t(A)$.
(ii) $\rho_t(aM) = 0$ *for every* $M \in \ell_1(A)$.

Proof. Indeed, (i) $\Longrightarrow$ (ii) follows from Corollary 3.24, and (ii) $\Longrightarrow$ (i) was proved in Theorem 3.21. □

We will prove now that $\mathcal{R}_t(A)$ is the *largest* tensor quasinilpotent ideal.

Theorem 3.26. *If* I *is a tensor quasinilpotent (possible, one-sided) ideal of* A *then* $I \subset \mathcal{R}_t(A)$.

Proof. Let I be a right ideal of A, and let $a \in I$. Then $aM \in \ell_1(I)$ for every $M \in \ell_1(A)$. As I is tensor quasinilpotent then $\rho_t(aM) = 0$ for every $M \in \ell_1(A)$. By Theorem 3.21, $a \in \mathcal{R}_t(A)$. So $I \subset \mathcal{R}_t(A)$.

If I is a left ideal of A and $a \in I$, then $\rho_t(aM) = \rho_t(Ma)$ by Proposition 3.3. So $\rho_t(aM) = 0$ for every $M \in \ell_1(A)$. We have again that $a \in \mathcal{R}_t(A)$. □

Lemma 3.27. *Let* $M = \{a_n\}_1^\infty$ *be a summable family in* A *and* $g : A \longrightarrow B$ *be a bounded homomorphism of normed algebras. Then* $g(M) := \{g(a_n)\}_1^\infty$ *is a summable family of* B *and* $\rho_t(g(M)) \leq \rho_t(M)$.

Proof. Indeed, it suffices to note that

$$\eta(g(M)^n) = \eta(g(M^n)) \leq \|g\| \, \eta(M^n)$$

for every n. □

Theorem 3.28. *Let* A *and* B *be a normed algebras, and let* $g : A \longrightarrow B$ *be an open bounded epimorphism. Then* $g(\mathcal{R}_t(A)) \subset \mathcal{R}_t(B)$.

Proof. Let $N = \{b_n\}_1^\infty$ be a summable family of B. As g is open, there is a summable family $M = \{a_n\}$ in A such that $g(M) = N$. It follows from Lemma 3.27 that $\rho_t(g(K)) \leq \rho_t(K)$ for every $K \in \ell_1(A)$. So if $a \in \mathcal{R}_t(A)$ then, by Corollary 3.25, we obtain that $\rho_t(g(a)N) \leq \rho_t(aM) = 0$ for an arbitrary $N \in \ell_1(B)$. Hence $g(a) \in \mathcal{R}_t(B)$ by Corollary 3.25. □

Let A be a normed algebra. Recall that if I is a closed ideal of A, then by a/I (and also by $q_I(a)$) we denote the element $a + I$ of the algebra A/I. By M/I we denote the family $\{a_n/I\}$, for every $M = \{a_n\} \in \ell_1(A)$.

Theorem 3.29. *Let* $M = \{a_n\}_1^\infty$ *be a summable family in a normed algebra* A. *Then* $\rho_t(M) = \rho_t(M/I)$ *for each closed tensor quasinilpotent ideal* I. *In particular,* $\rho_t(M) = \rho_t(M/\mathcal{R}_t(A))$.

Proof. As clearly $\rho_t(M/\mathcal{R}_t(A)) \leq \rho_t(M)$, it suffices to show the reverse inequality. Let $\delta > 0$, $n \in \mathbb{N}$ and $M^n = \{b_m\}_1^\infty$. Then for every m there are $c_m \in A$ and $d_m \in I$ such that $b_m = c_m + d_m$ and

$$\|c_m\| \leq \|b_m/I\| + 2^{-m}\delta.$$

Let $N = \{c_m\}_1^\infty$ and $S = \{d_m\}_1^\infty$. Since $\eta(N) \leq \eta(M^n/I) + \delta$, then $N \in \ell_1(A)$. As $S = M^n - N$, we have that $S \in \ell_1(I)$ and that

$$\rho_t(N + S) \leq \rho_t(N \sqcup S)$$

by Proposition 3.3. As $I \subset \mathcal{R}_t(A)$, we have that

$$\rho_t(N + S) \leq \rho_t(N \sqcup S) = \rho_t(N).$$

by Theorem 3.22. Therefore we obtain that

$$\rho_t(M)^n \overset{(3.5)}{=} \rho_t(M^n) = \rho_t(N + S) \leq \rho_t(N) \leq \eta(N) \leq \eta(M^n/I) + \delta.$$

Since δ is arbitrary, then $\rho_t(M)^n \leq \eta(M^n/I)$ for every $n > 0$. Taking n-roots and passing to limits, we obtain that $\rho_t(M) \leq \rho_t(M/I)$. □

Corollary 3.30. *Let A be a normed algebra. Then $\mathcal{R}_t(A/\mathcal{R}_t(A)) = 0$.*

Proof. Let $a/\mathcal{R}_t(A) \in \mathcal{R}_t(A/\mathcal{R}_t(A))$. Then it follows from Theorem 3.29 that $\rho_t(\{a\} \sqcup M) = \rho_t(M)$ for every $M \in \ell_1(A)$. Hence $a \in \mathcal{R}_t(A)$, and therefore $\mathcal{R}_t(A/\mathcal{R}_t(A)) = 0$. □

Theorem 3.31. *Let A be a normed algebra. If I is an ideal of A then $\mathcal{R}_t(I) = \mathcal{R}_t(A) \cap I$.*

Proof. It is clear that $\mathcal{R}_t(A) \cap I \subset \mathcal{R}_t(I)$.

Let $a \in \mathcal{R}_t(I)$. For every $M \in \ell_1(A)$, we have that $MaM \in \ell_1(I)$ and then

$$\rho_t(aM)^2 = \rho_t(aMaM) = 0$$

by (3.5) and Corollary 3.25. Therefore $a \in \mathcal{R}_t(A)$ and $\mathcal{R}_t(I) \subset \mathcal{R}_t(A) \cap I$. □

Note that this result contains Lemma 3.18 and implies that

$$\mathcal{R}_t(\mathcal{R}_t(A)) = \mathcal{R}_t(A)$$

for every normed algebra A.

3.6. Tensor radical algebras and ideals

A normed algebra A is called *tensor radical* if the projective tensor product $A\widehat{\otimes}_\gamma B$ is radical for every normed algebra B. It is evident that A is tensor radical if and only if its completion $\widehat{A}$ is tensor radical. If A is tensor radical then its opposite algebra A^{op} is also tensor radical. An ideal of a normed algebra is called *tensor radical* if it is a tensor radical algebra.

The following result is an easy consequence of associativity and distributivity of tensor product.

Proposition 3.32. *Let A and B be normed algebras.*

(i) *If A is tensor radical then $A\widehat{\otimes}B$ is tensor radical.*
(ii) *If A and B are tensor radical then $A \oplus B$ is tensor radical.*

The study of deeper properties is based on the following theorem.

Theorem 3.33. *For a normed algebra A the following conditions are equivalent.*

(i) *A is tensor radical.*
(ii) *A is tensor quasinilpotent.*

Proof. (ii) $\Rightarrow$ (i) follows from Corollary 3.5, taking into account that every element of $A\widehat{\otimes}B$ can be represented as $M_{\otimes}L$ for some $M \in \ell_1(A)$ and $L \in \ell_\infty(B)$.
(i) $\Rightarrow$ (ii) follows from Theorem 3.1. Indeed, by this theorem, there are a Banach algebra B and $L \in \ell_\infty(B)$ such that $\rho(M_{\otimes}L) = \rho_t(M)$ for every $M \in \ell_1(A)$. If A is tensor radical then $A\widehat{\otimes}B$ is radical, whence $\rho(M_{\otimes}L) = 0$ for every $M \in \ell_1(A)$. Then $\rho_t(M) = 0$ for every $M \in \ell_1(A)$, i.e., A is tensor quasinilpotent. □

Corollary 3.34. *Every subalgebra of a tensor radical normed algebra is tensor radical.*

Proof. Follows from Theorem 3.33, since subalgebras of a tensor radical algebra are obviously tensor quasinilpotent. □

Corollary 3.35. *Let A be a normed algebra. If there is a tensor quasinilpotent dense subalgebra B of A then A is tensor quasinilpotent.*

Proof. Indeed, as B is tensor radical by Theorem 3.33, the completion $\widehat{B}$ is also tensor radical. As $\widehat{B}$ and $\widehat{A}$ are identified, the algebra $\widehat{A}$ is tensor quasinilpotent by Theorem 3.33. Therefore A is tensor quasinilpotent. □

As a consequence of Corollary 3.23 and Theorem 3.33, for every normed algebra A, $\mathcal{R}_t(A)$ is the largest tensor radical ideal of A.

Theorem 3.36. *Let A be a normed algebra and $a \in A$. Then $a \in \mathcal{R}_t(A)$ if and only if $a \otimes b \in \mathrm{Rad}(A\widehat{\otimes}B)$, for every normed algebra B and $b \in B$.*

Proof. Let $a \in \mathcal{R}_t(A)$. Then $a \otimes b \in \mathcal{R}_t(A)\widehat{\otimes}^{(\cdot)}B$ for an arbitrary normed algebra B and for every $b \in B$. In the same time, $\mathcal{R}_t(A)\widehat{\otimes}B$ is a radical Banach algebra. Being the image of a bounded homomorphism from $\mathcal{R}_t(A)\widehat{\otimes}B$, $\mathcal{R}_t(A)\widehat{\otimes}^{(\cdot)}B$ consists of quasinilpotent elements of $A\widehat{\otimes}B$. But it is also an ideal of $A\widehat{\otimes}B$. So $\mathcal{R}_t(A)\widehat{\otimes}^{(\cdot)}B \subset \mathrm{Rad}(A\widehat{\otimes}B)$ and therefore $a \otimes b \in \mathrm{Rad}(A\widehat{\otimes}B)$.

Suppose that $a \otimes b \in \mathrm{Rad}(A\widehat{\otimes}B)$ for every normed algebra B and $b \in B$. Take B as in Theorem 3.1(ii). Then B has the identity element 1 and there is a family $L \in \ell_\infty(B)$ such that $\rho(M_{\otimes}L) = \rho_t(M)$ for every $M \in \ell_1(A)$. Since $a \otimes 1 \in \mathrm{Rad}(A\widehat{\otimes}B)$ then $\rho((a \otimes 1)(M_{\otimes}L)) = 0$. As $(a \otimes 1)(M_{\otimes}L) = aM_{\otimes}L$, we have that $\rho_t(aM) = 0$ for every $M \in \ell_1(A)$. By Theorem 3.21, $a \in \mathcal{R}_t(A)$. □

Proposition 3.37. *Let A and B be normed algebras. Then*

$$\mathcal{R}_t(A)\widehat{\otimes}^{(\cdot)}B \subset \mathcal{R}_t(A\widehat{\otimes}B) \subset \mathrm{Rad}(A\widehat{\otimes}B).$$

Proof. If $a \in \mathcal{R}_t(A)$ then it follows from Theorem 3.36 that

$$a \otimes b \otimes c \in \operatorname{Rad}\left(A\widehat{\otimes}B\widehat{\otimes}C\right),$$

for every $b \in B$ and for every normed algebra C and $c \in C$. By the same theorem, $a \otimes b \in \mathcal{R}_t\left(A\widehat{\otimes}B\right)$ for every $b \in B$. So the closure of $\mathcal{R}_t(A) \otimes B$ in $A\widehat{\otimes}B$ is contained into $\mathcal{R}_t\left(A\widehat{\otimes}B\right)$ since $\mathcal{R}_t\left(A\widehat{\otimes}B\right)$ is closed. But $\mathcal{R}_t(A)\widehat{\otimes}^{(\cdot)}B$ is generated as a normed algebra by elements of $\mathcal{R}_t(A) \otimes B$. Hence $\mathcal{R}_t(A)\widehat{\otimes}^{(\cdot)}B \subset \mathcal{R}_t\left(A\widehat{\otimes}B\right)$. As $A\widehat{\otimes}B$ is a Banach algebra, $\mathcal{R}_t\left(A\widehat{\otimes}B\right) \subset \operatorname{Rad}\left(A\widehat{\otimes}B\right)$ by Corollary 3.20. □

Proposition 3.38. *Let A be a normed algebra and I be an ideal of A. If I and $A/\overline{I}$ are tensor radical then A is tensor radical.*

Proof. Let $M \in \ell_1(A)$. As I is tensor radical then, as we mentioned above, the closure $\overline{I}$ of I in A is also tensor radical. Then $\rho_t(M) = \rho_t\left(M/\overline{I}\right)$ by Theorem 3.29, but $\rho_t\left(M/\overline{I}\right) = 0$ by Theorem 3.33. So A is tensor radical. □

Proposition 3.39. *Let A be a normed algebra and I be a flexible ideal of A. If A is tensor radical then $(I, \|\cdot\|_I)$ is tensor radical.*

Proof. Let $M = (a_n)_{n=1}^{\infty}$ be a summable family in I. Then M is summable in A and

$$\begin{aligned}\eta_{\|\cdot\|_A}(M^n) \le \eta_{\|\cdot\|_I}(M^n) &= \sum_{i_1,\dots,i_n} \left\|a_{i_1}\cdots a_{i_{n-1}}a_{i_n}\right\|_I \\ &\le \sum_{i_1,\dots,i_{n-1}} \left\|a_{i_1}\cdots a_{i_{n-1}}\right\|_A \sum_{i_n} \|a_{i_n}\|_I \le \eta_{\|\cdot\|_A}\left(M^{n-1}\right)\eta_{\|\cdot\|_I}(M).\end{aligned}$$

Thus $\rho_{t|I}(M) = \rho_t(M) = 0$, where $\rho_{t|I}$ is the tensor spectral radius in $(I, \|\cdot\|_I)$. □

Corollary 3.40. *Let I and J be flexible ideals of a normed algebra A. If I and J are tensor radical then $I \cap J$ and $I + J$ are tensor radical with respect to their flexible norms (see Proposition 2.7).*

Proof. By Corollary 3.25, I and J are contained in $\mathcal{R}_t(A)$. Hence the same is true for the ideals $I \cap J$ and $I + J$. It follows that they are tensor radical with respect to the norm $\|\cdot\|_A$. By Proposition 3.39, they are tensor radical with respect to their flexible norms. □

The following result will be often applied in the further sections.

Lemma 3.41. *Let A_1, A_2 be normed algebras, $A = A_1\widehat{\otimes}A_2$, and let $J_i \subset I_i$ be ideals of A_i, for $i = 1, 2$. Denote by J the ideal of A generated by $J_1 \otimes A_2 + A_1 \otimes J_2$, and let I be the ideal of A generated by $I_1 \otimes A_2 + A_1 \otimes I_2$. If $\overline{I_i/J_i}$ are tensor radical then $\overline{I} \subset Q_J(A)$.*

Proof. Let π be a strictly irreducible representation of A such that $\pi(J) = 0$. First we show that $\pi(I_1 \otimes A_2) = 0$ and $\pi(A_1 \otimes I_2) = 0$.

Assume, to the contrary, that $\pi(I_1 \otimes A_2) \neq 0$. Hence the restriction τ of π to $\overline{I_1}\widehat{\otimes}^{(\cdot)} A_2$ is strictly irreducible. As $\tau(J_1 \otimes A_2) = 0$ then $\tau\left(\overline{J_1} \otimes A_2\right) = 0$ because one may assume that τ is continuous.

Moreover, τ induces a strictly irreducible representation of the algebra $C := (\overline{I_1}/\overline{J_1})\widehat{\otimes} A_2$, because the composition of natural maps

$$(\overline{I_1}/\overline{J_1})\widehat{\otimes} A_2 \longrightarrow \left(\overline{I_1}\widehat{\otimes} A_2\right)/\overline{J_1 \otimes A_2} \longrightarrow \left(\overline{I_1}\widehat{\otimes}^{(\cdot)} A_2\right)/\overline{J_1 \otimes A_2}'$$

is a contractive epimorphism, where $\overline{J_1 \otimes A_2}$ and $\overline{J_1 \otimes A_2}'$ are the closures of $J_1 \otimes A_2$ in $\overline{I_1}\widehat{\otimes} A_2$ and $\overline{I_1}\widehat{\otimes}^{(\cdot)} A_2$, respectively, and clearly $\overline{J_1 \otimes A_2} \subset \overline{J_1 \otimes A_2}'$.

As there exists a non-zero strictly irreducible representation of C then C is not radical, but C is radical in virtue of tensor radicality of $\overline{I_1}/\overline{J_1}$, a contradiction. Hence $\pi(I_1 \otimes A_2) = 0$ and, similarly, $\pi(A_1 \otimes I_2) = 0$. As I lies in the intersection of all primitive ideals of A containing J, so does $\overline{I}$. By Proposition 2.1(i), $\overline{I} \subset Q_J(A)$. □

3.7. Algebras commutative modulo the tensor radical

We say that a normed algebra A is *commutative modulo the tensor radical* if the algebra $A/\mathcal{R}_t(A)$ is commutative. An equivalent condition is $[a, b] \in \mathcal{R}_t(A)$ for all $a, b \in A$.

Theorem 3.42. *If normed algebras A_1 and A_2 are commutative modulo the tensor radical then the same is true for $A := A_1 \widehat{\otimes} A_2$.*

Proof. By Proposition 3.37, $\mathcal{R}_t(A_1)\widehat{\otimes} A_2 + A_1 \widehat{\otimes}^{(\cdot)} \mathcal{R}_t(A_2) \subset \mathcal{R}_t(A)$. Then

$$[a_1 \otimes b_1, a_1 \otimes b_2] = [a_1, a_2] \otimes b_1 b_2 + a_2 a_1 \otimes [b_1, b_2] \in \mathcal{R}_t(A)$$

for all $a_1, a_2 \in A_1$ and $b_1, b_2 \in A_2$. Hence $[c_1, c_2] \in \mathcal{R}_t(A)$ for all $c_1, c_2 \in A$. □

Theorem 3.43. *Let A_1 be a normed algebra, and let A_2 be a Banach algebra. If A_1 is commutative modulo the tensor radical and A_2 is radical then $A_1 \widehat{\otimes} A_2$ is radical.*

Proof. Let $A = A_1 \widehat{\otimes} A_2$ and $I = \mathcal{R}_t(A_1)\widehat{\otimes}^{(\cdot)} A_2$. The Banach ideal I of A is radical, being isometric to the quotient of the radical algebra $\mathcal{R}_t(A_1)\widehat{\otimes} A_2$. On the other hand, the quotient $A/\overline{I}$ is topologically isomorphic to $(A/\mathcal{R}_t(A_1))\widehat{\otimes} A_2$ by Proposition 2.11. Since $A/\mathcal{R}_t(A_1)$ is commutative and A_2 is radical, the algebra $(A/\mathcal{R}_t(A_1))\widehat{\otimes} A_2$ is radical by [2, Theorem 4.4.2]. Thus $A/\overline{I}$ is radical, and A is radical by Lemma 2.2. □

3.8. Relation with joint spectral radius

In 1960 Rota and Strang [21] defined a notion of spectral radius for bounded subsets of a Banach algebra. This definition holds for normed algebras (and we already introduced it for countable bounded subsets in Section 3.1). Namely, if K

is a bounded subset of a normed algebra A then its *joint spectral radius* $\rho(K)$ is defined by

$$\rho(K) = \inf_{n\in\mathbb{N}} \|K^n\|^{1/n},$$

where the norm of a set is defined as the supremum of the norms of its elements, and the products of sets are defined by $KN = \{ab : a \in K, b \in N\}$. Since $\|K^{n+k}\| \le \|K^n\|\|K^k\|$ for every $n, k > 0$, one has that

$$\rho(K) = \lim_{n\to\infty} \|K^n\|^{1/n}. \tag{3.11}$$

Taking $n = mk$ in (3.11) for $m = 1, 2, \ldots$, we observe that

$$\rho(K^k) = \rho(K)^k$$

for every bounded $K \subset A$ and integer $k > 0$.

It was proved in [25, Theorem 3.5] that the joint spectral radius is a subharmonic function. This means that if $\lambda \to K(\lambda)$ is an analytic map in a natural sense from a domain $\mathcal{D} \subset \mathbb{C}$ into the set of bounded subsets of A then the function $\lambda \to \rho(K(\lambda))$ is subharmonic.

Let K be a subset of A, and let $\mathrm{F}_\infty(K)$ be the set of all bounded families $N = \{a_n\}_1^\infty$ with $a_n \in K$ for every $n > 0$. Clearly $\mathrm{F}_\infty(K)$ is a metric space with respect to the metric induced by the norm of $\ell_\infty(A)$. In particular, we have that $\mathrm{F}_\infty(A) = \ell_\infty(A)$. If K is bounded, it follows from [25, Proposition 2.2] that there is a family $L = \{b_n\}_1^\infty \in \mathrm{F}_\infty(K)$ such that $\rho(K) = \rho(L)$. So

$$\rho(K) = \max_{N\in \mathrm{F}_\infty(K)} \rho(N)$$

and this allows to obtain some results on joint spectral radius of bounded subsets considering bounded families.

The following property is important for our applications: *If $\rho(K) = 0$ then the linear span of K consists of quasinilpotent elements.* This result from [24] can be easily proved by the direct evaluation of the norms of powers for $\sum_{i=1}^n \lambda_i a_i$, where $a_i \in K$. The following result is similar.

Lemma 3.44. *Let K be a bounded subset of a normed algebra A. If $\rho(K) = 0$ then $\rho(\sum_{n=1}^\infty \lambda_n a_n) = 0$ for each sequence $a_n \in K$ and each summable sequence λ_n of complex numbers, where $\sum_{n=1}^\infty \lambda_n a_n$ are elements of the completion $\widehat{A}$ as usual.*

Proof. Let $M = \{\lambda_n\}_1^\infty$ and $L = \{a_n\}_1^\infty$. As $\mathbb{C}\widehat{\otimes}A$ is identified with $\widehat{A}$ under the identification $\lambda\widehat{\otimes}x$ with λx, we obtain that

$$\rho\left(\sum_{n=1}^\infty \lambda_n a_n\right) \le \rho_t(M)\,\rho(L) \le \rho_t(M)\,\rho(K) = 0$$

by Theorem 3.1(i). □

We say that a normed algebra A is *compactly quasinilpotent* [29] if $\rho(K) = 0$ for each precompact subset K of A. The following statement improves [29, Theorem 4.29] which was proved for Banach algebras.

Theorem 3.45. *Every compactly quasinilpotent normed algebra is tensor radical.*

Proof. Let A be a compactly quasinilpotent normed algebra, B a Banach algebra, and let $x = \sum_{n=1}^{\infty} a_n \otimes b_n \in A\widehat{\otimes}B$. One can assume that $\{a_n\}_1^{\infty}$ consists of elements of A (see Section 2.4.1), the sequence $\alpha_n = \|a_n\|$ is summable, while $\|b_n\| \leq 1$ for every n. It is obvious that there exists a sequence $\varepsilon_n \to 0$ such that $\lambda_n := \alpha_n/\varepsilon_n$ is summable. Set $c_n = \lambda_n^{-1} a_n$ for $n > 0$. Then $\|c_n\| \to 0$ as $n \to \infty$, so the set $N := \{c_n : n = 1, 2, \ldots\}$ is precompact and $\rho(N) = 0$ by our assumption. For $K = \{c_n \otimes b_n : n = 1, 2, \ldots\}$, it is easy to check that

$$\|K^k\| \leq \|N^k\|,$$

whence $\rho(K) \leq \rho(N) = 0$. Applying Lemma 3.44, we obtain that

$$\rho(x) = \rho\left(\sum_{n=1}^{\infty} \lambda_n c_n \otimes b_n\right) = 0.$$

Therefore $A\widehat{\otimes}B$ consists of quasinilpotent elements. □

It was proved in [29] that each normed algebra has the largest compactly quasinilpotent ideal $\mathcal{R}_c(A)$.

Corollary 3.46. *Let A be a normed algebra. Then $\mathcal{R}_c(A) \subset \mathcal{R}_t(A) \subset A \cap \mathrm{Rad}(\widehat{A})$.*

Proof. The first inclusion follows by Theorem 3.45.

By Corollary 3.35, the closure $\overline{\mathcal{R}_t(A)}$ in the completion $\widehat{A}$ of A is tensor quasinilpotent. Then $\overline{\mathcal{R}_t(A)}$ consists of quasinilpotent elements of $\widehat{A}$. As $\overline{\mathcal{R}_t(A)}$ is an ideal of $\widehat{A}$ consisting of quasinilpotents, $\overline{\mathcal{R}_t(A)} \subset \mathrm{Rad}(\widehat{A})$. Therefore we obtain that $\mathcal{R}_t(A) \subset A \cap \mathrm{Rad}(\widehat{A})$. □

3.9. Compactness conditions

It is still an open problem if any radical Banach algebra is tensor radical. We will show here that the answer is positive if A has some compactness properties.

In the well-known paper of Vala [32] it was shown that

(i) For compact operators a, b on a Banach space X the multiplication operator $x \longmapsto axb$ is compact on $\mathcal{B}(X)$.

(ii) If the operator $x \longmapsto axa$ is compact then the operator a is compact.

This gave a possibility to introduce a notion of a compact element of a normed algebra: an element a of A is *compact* if

$$W_a := L_a R_a$$

is a compact operator, where L_a and R_a are defined by $L_a x = ax$ and $R_a x = xa$ for every $x \in A$. Similarly, one says that a is a *finite rank element* of A if W_a has finite rank. Basing on these definitions there were introduced various Banach-algebraic analogues of the class of algebras of compact operators. The most popular one is the class of compact algebras: A is *compact* if all its elements are compact. Slightly more narrow but much more convenient is the class of bicompact algebras:

A is *bicompact* if $L_a R_b$ are compact for all $a, b \in A$. Furthermore, A is called an *approximable algebra* if the set of finite rank elements is dense in A. These classes are closed under passing to ideals and quotients, but in general not stable under extensions.

To overcome this obstacle and considerably extend the class of algebras in consideration, let us say that a normed algebra A is *hypocompact* (respectively, *hypofinite*) if each non-zero quotient of A has a non-zero compact (respectively, finite rank) element. It is not difficult to check that the class of all hypocompact algebras is closed under extensions, as well as under passing to ideals and quotients. It is not known if it is closed under passing to subalgebras. One can realize a hypocompact algebra as a result of a transfinite sequence of extensions of bicompact algebras, but we will need a close result, see Proposition 3.48 below.

Since a quotient of a quotient of A is isomorphic to a quotient of A, the following result is an immediate consequence of the definition of hypocompact algebras.

Corollary 3.47. *A quotient of a hypocompact normed algebra (by a closed ideal) is hypocompact.*

A similar result is valid for hypofinite normed algebras. We also need the following result.

Proposition 3.48. *Let A be a normed algebra. Then A is hypocompact (respectively, hypofinite) if and only if there is an increasing transfinite chain $(J_\alpha)_{\alpha\leq\beta}$ of closed ideals of A such that $J_0 = 0$, $J_\beta = A$, $J_\alpha = \overline{\cup_{\alpha'<\alpha} J_{\alpha'}}$ for every limit ordinal $\alpha \leq \beta$ and $J_{\alpha+1}/J_\alpha$ is a non-zero ideal of A/J_α having a dense set of compact (respectively, finite rank) elements of A/J_α for every ordinal α between 0 and β.*

Proof. $\Rightarrow$ Let us use the transfinite induction. Let $J_0 = 0$. If we constructed J_α and $J_\alpha \neq A$ then take a non-zero compact (finite rank) element b in A/J_α and denote by K the closed ideal of A/J_α generated by b. Let us define $J_{\alpha+1}$ as the preimage of K in A:

$$J_{\alpha+1} = \{c \in A : c/J_\alpha \in K\}.$$

Then $J_{\alpha+1}/J_\alpha$ is topologically isomorphic to K. It remains to note that the chain is stabilized at some step β because A has a definite cardinality. So $J_\beta = A$.

$\Leftarrow$ Let I be a closed ideal of A and $I \neq A$. Then there is the first ordinal $\alpha' < \beta$ such that I is not contained into $J_{\alpha'}$. Then $J_\alpha \subset I$ for every $\alpha < \alpha'$, whence $\overline{\cup_{\alpha<\alpha'} J_\alpha} \subset I$. So α' has a precessor α'': $\alpha' = \alpha'' + 1$. Take $a \in J_{\alpha'} \backslash I$, and let $G = \{x \in A : \|x - a\| < \text{dist}\,(a, I)\}$. Then $G/J_{\alpha''} := \{x/J_{\alpha''} \in A/J_{\alpha''} : x \in G\}$ is an open neighbourhood of $a/J_{\alpha''} \in J_{\alpha'}/J_{\alpha''}$ and therefore has a compact (finite rank) element $b/J_{\alpha''}$ of $A/J_{\alpha''}$. It is clear that b/I is a non-zero compact (finite rank) element of A/I. So A is hypocompact (respectively, hypofinite). $\square$

Corollary 3.49. *Let B be a normed algebra, and let A be a hypocompact (respectively, hypofinite) dense subalgebra of B. Then B is hypocompact (respectively, hypofinite).*

Proof. let $(J_\alpha)_{\alpha\leq\beta}$ be a transfinite chain of ideals of A described in Proposition 3.48. Let $I_\alpha = \overline{J_\alpha}$, the closure of J_α in B, for every ordinal $\alpha \leq \beta$. Then the chain $(I_\alpha)_{\alpha\leq\beta}$ satisfies the conditions of Proposition 3.48. So B is hypocompact (respectively, hypofinite) by Proposition 3.48. □

The following result of [26] will be very useful in Section 5.

Theorem 3.50. *If a Banach algebra is hypocompact then spectra of its elements are (finite or) countable.*

Our main aim here is to show that for hypocompact Banach algebras the ideal $\mathcal{R}_t(A)$ coincides with the Jacobson radical $\mathrm{Rad}(A)$.

For a bounded subset M of a Banach algebra, set

$$r(M) = \limsup_{n\to\infty} \left(\sup\{\rho(a) : a \in M^n\}\right)^{1/n}.$$

Clearly $r(M) \leq \rho(M)$.

This spectral characteristic, introduced (for sets of matrices) in 1992 by M.A. Berger and Y. Wang [7], turned out to be very useful in operator theory. It was proved in [7] that $r(M) = \rho(M)$ for any bounded set M of matrices. In [25] the authors showed that the same is true if M is a precompact set of compact operators on a Banach space. In the further works [26, 27, 30] there were obtained several extensions of this result. Here we need the following consequence of [30, Theorem 4.11] (where only Banach algebras were considered).

Corollary 3.51. *Let A be a hypocompact normed algebra. Then $r(M) = \rho(M)$ for each precompact subset M of A.*

Proof. It is clear that $r(M)$ and $\rho(M)$ don't change if pass to the completion $\widehat{A}$. Moreover, $\widehat{A}$ is hypocompact by Corollary 3.49. Now, applying [30, Theorem 4.11], we get the result. □

Theorem 3.52. *If A is a hypocompact normed algebra then*

$$\mathcal{R}_c(A) = \mathcal{R}_t(A) = A \cap \mathrm{Rad}(\widehat{A}).$$

Proof. Taking into account Corollary 3.46, we have to prove only the inclusion $A\cap\mathrm{Rad}(\widehat{A}) \subset \mathcal{R}_c(A)$. Since $A\cap\mathrm{Rad}(\widehat{A})$ is an ideal of a hypocompact algebra A, it is hypocompact (see Corollary 3.60). If M is a precompact subset of $A\cap\mathrm{Rad}(\widehat{A})$ then $r(M) = 0$ because all elements in $\cup_{n=1}^{\infty} M^n$ are quasinilpotent. By Corollary 3.51, $\rho(M) = 0$. So $A \cap \mathrm{Rad}(\widehat{A})$ is a compactly quasinilpotent ideal of A which implies that $A \cap \mathrm{Rad}(\widehat{A}) \subset \mathcal{R}_c(A)$. □

Corollary 3.53. *Each radical hypocompact normed algebra is tensor radical.*

The following result [1, Corollary 6.2] supplies us with an important class of examples of bicompact radical Banach algebras. Our proof of radicality of $\mathcal{K}(X)/\mathcal{A}(X)$ differs from the proof in [1].

Lemma 3.54. *Let $\mathcal{K}(X)$ be the algebra of all compact operators on a Banach space X, $\mathcal{A}(X)$ be the closure in $\mathcal{K}(X)$ of the ideal $\mathcal{F}(X)$ of finite rank operators. Then $\mathcal{K}(X)/\mathcal{A}(X)$ is a radical bicompact Banach algebra.*

Proof. The fact that $\mathcal{K}(X)$ is bicompact follows from the mentioned result of Vala [32]. Since the quotient of a bicompact algebra is obviously bicompact, $\mathcal{K}(X)/\mathcal{A}(X)$ is bicompact.

To see that $\mathcal{K}(X)/\mathcal{A}(X)$ is radical, note that all projections in $\mathcal{K}(X)$ are of finite rank and therefore belong to $\mathcal{A}(X)$. It follows that for each $a \in \mathcal{K}(X)$ and each spectral (= Riesz) projection p of a corresponding to a subset $\alpha \subset \mathbb{C}$ non-containing 0, we have that

$$q((1-p)a) = q(a),$$

where q is the quotient map from $\mathcal{K}(X)$ to $\mathcal{K}(X)/\mathcal{A}(X)$. Since

$$\rho(q((1-p)a)) \leq \rho((1-p)a)$$

and $\rho((1-p)a)$ can be made arbitrary small by an appropriate choice of p, we conclude that $\rho(q(a)) = 0$. So $\mathcal{K}(X)/\mathcal{A}(X)$ consists of quasinilpotent elements. □

Recall that *Riesz operators* are defined [23] as operators that are quasinilpotent modulo the compact operators.

Corollary 3.55. *For every Riesz operator $a \in \mathcal{B}(X)$ and for every $\varepsilon > 0$, there is $m \in \mathbb{N}$ such that $\operatorname{dist}_{\|\cdot\|_{\mathcal{B}}}(a^n, \mathcal{F}(X)) < \varepsilon^n$ for each $n > m$.*

Proof. Indeed, it follows from Lemma 3.54 that every Riesz operator is quasinilpotent modulo the approximable operators. □

Our aim now is to show that the class of hypocompact algebras is stable under tensor products. Let ball(A) denote the closed unit ball of A. Recall that $W_a = L_a R_a$ for every $a \in A$, and if $M \subset A$ and $N \subset B$ are not subspaces then $M \otimes N$ means only the set $\{a \otimes b : a \in M, b \in N\}$, not its linear span. Moreover, if I is a closed ideal of A, then M/I means the set $\{a/I : a \in M\} \subset A/I$.

Lemma 3.56. *Let A, B be unital Banach algebras, J a closed ideal in $A \hat{\otimes} B$, I_1 and I_2 closed ideals in A and B respectively. Let elements $a \in A$, $b \in B$ satisfy the conditions $a \otimes I_2 \subset J$ and $I_1 \otimes b \subset J$. If a/I_1 and b/I_2 are compact elements of A/I_1 and B/I_2 respectively then $(a \otimes b)/J$ is a compact element of $(A \widehat{\otimes} B)/J$.*

Proof. Note that $aAa \otimes I_2 = (a \otimes I_2)(Aa \otimes 1) \subset J$ and, similarly, $I_1 \otimes bBb \subset J$. Choose $\varepsilon > 0$. Let $x_1, \ldots, x_n \in$ ball(A) be such that $\{W_a x_i / I_1 : 1 \leq i \leq n\}$ is an ε-net in $W_a(\text{ball}(A/I_1))$. This means that for any $x \in$ ball(A) there are $i \leq n$ and $x' \in I_1$ with

$$\|W_a x - W_a x_i - x'\| < \varepsilon.$$

In the same way one finds $y_1, \ldots, y_m \in$ ball(B) such that for each $y \in$ ball(B) there are $k \leq m$ and $y' \in I_2$ with

$$\|W_b y - W_b y_k - y'\| < \varepsilon.$$

Let us check that the set $\{(W_a x_i \otimes W_b y_k)/J : i \le n,\ k \le m\}$ is a δ-net for the set $(W_a(\mathrm{ball}(A)) \otimes W_b(\mathrm{ball}(B)))/J$, where $\delta = (\|a\|^2 + \|b\|^2)\varepsilon$. Indeed, for $x \in \mathrm{ball}(A)$, $y \in \mathrm{ball}(B)$, choose i, k as above. Then we obtain that

$$\begin{aligned} z &:= W_a x \otimes W_b y - W_a x_i \otimes W_b y_k \\ &= (W_a x - W_a x_i) \otimes W_b y + W_a x_i \otimes (W_b y - W_b y_k) \\ &= x' \otimes W_b y + W_a x_i \otimes y' + u \otimes W_b y + W_a x_i \otimes v \end{aligned}$$

where $\|u\| < \varepsilon$, $\|v\| < \varepsilon$. Since the first two summands belong to J, we conclude that the norm of z/J in $A \widehat{\otimes} B/J$ is less than δ.

We proved that $(W_a(\mathrm{ball}(A)) \otimes W_b(\mathrm{ball}(B)))/J$ is precompact in $\left(A \widehat{\otimes} B\right)/J$. Since $\mathrm{ball}(A \widehat{\otimes} B)$ is the closed convex hull of the set $\mathrm{ball}(A) \otimes \mathrm{ball}(B)$, then the set $(W_a \otimes W_b)(\mathrm{ball}(A \widehat{\otimes} B))/J$ is precompact, whence $W_{(a \otimes b)/J}$ is compact. □

Let us say, for brevity, that an element a is *compact modulo a closed ideal* J if a/J is a compact element of A/J.

Theorem 3.57. *If normed algebras A and B are hypocompact then $A \widehat{\otimes} B$ is hypocompact.*

Proof. By Corollary 3.49, one can assume that A and B are complete.

Suppose first that A and B are unital. Let J be a proper closed ideal of $A \widehat{\otimes} B$. We have to prove that $\left(A \widehat{\otimes} B\right)/J$ has non-zero compact elements.

Set $I_1 = \{x \in A : x \otimes B \subset J\}$. By our assumption, $I_1 \neq A$ (indeed, otherwise $J = A \widehat{\otimes} B$), so there exists an element $a \in A \backslash I_1$ which is compact modulo I_1. Set $I_2 = \{y \in B : a \otimes y \in J\}$. Since $a \notin I_1$ then $I_2 \neq B$. Let $b \in B \backslash I_2$ be an element of B compact modulo I_2.

By the definition of I_2, $a \otimes I_2 \subset J$. Furthermore, $I_1 \otimes b \subset J$ because $I_1 \otimes B \subset J$. Hence the assumptions of Lemma 3.56 are satisfied, therefore $a \otimes b$ is an element of $A \widehat{\otimes} B$ compact modulo J. It is clear that $a \otimes b \notin J$ by the choice of b.

In general it suffices to note that the unitization A^1 of a hypocompact algebra A is hypocompact and $A \widehat{\otimes} B$, being a closed ideal of $A^1 \widehat{\otimes} B^1$, is hypocompact by Corollary 3.60. □

3.10. Topological radicals

A map P which associates with every normed algebra A a closed ideal $P(A)$ of A is called a *topological radical* if $P(P(A)) = P(A)$, $P(A/P(A)) = 0$, $P(I)$ is an ideal of A and $P(I) \subset P(A)$ for every ideal I of A, and $f(P(A)) \subset P(B)$ for every morphism $f : A \longrightarrow B$. The meaning of the later requirement depends on the specification of morphisms in the different categories whose objects are normed algebras. In the applications below, *open bounded epimorphisms* are included in the class of morphisms of any such category. The study of topological radicals was initiated by [13].

A topological radical P is called *hereditary* if $P(I) = I \cap P(A)$ for any ideal I of each normed algebra A. A normed algebra A is called *P-radical* if $A = P(A)$ and *P-semisimple* if $P(A) = 0$.

Note [13] that the Jacobson radical rad : $A \longmapsto \mathrm{rad}(A)$ is not a topological radical on the class of all normed algebras, but its restriction Rad to the class of all Banach algebras is a hereditary topological radical. This radical admits different extensions to the class of all normed algebras which are topological radicals (see for instance [29, Section 2.6]). One of them is the regular extension $\mathrm{Rad}^r : A \longmapsto A \cap \mathrm{Rad}\left(\widehat{A}\right)$ (see [29, Section 2.8]) where $\widehat{A}$ is the completion of A. We already met this radical in Corollary 3.46 and Theorem 3.52.

Let $\mathcal{R}_t$ denote the map $A \longmapsto \mathcal{R}_t(A)$ for every normed algebra A.

Theorem 3.58. *$\mathcal{R}_t$ is a hereditary topological radical in the category of normed algebras morphisms of which are open bounded epimorphisms.*

Proof. It follows from the results of Sections 3.4 and 3.5. □

The same was proved for the map $\mathcal{R}_c : A \longmapsto \mathcal{R}_c(A)$ (see [29, Theorem 4.25]).

Moreover, it was proved in [30, Theorem 3.14] (see also the short communication [26]) that each normed algebra A has a largest hypocompact ideal $\mathcal{R}_{hc}(A)$, and the map $\mathcal{R}_{hc} : A \longmapsto \mathcal{R}_{hc}(A)$ *is a hereditary topological radical on the class of normed algebras with open bounded epimorphisms as morphisms.* It should be noted that for simplicity the results of [30, Section 3.2] were formulated for Banach algebras, but the proofs did not use the completeness.

Theorem 3.59. *For every normed algebra A, there exists a largest hypofinite ideal $\mathcal{R}_{hf}(A)$, and the map $\mathcal{R}_{hf} : A \longmapsto \mathcal{R}_{hf}(A)$ is a hereditary topological radical on the class of normed algebras morphisms of which are bounded homomorphisms with dense image.*

Proof. Similar to the proof in [30, Section 3.2]. One have to replace compact algebras by approximable ones and take into account that a bounded homomorphism with dense image maps finite rank elements into finite rank elements. □

We note that $\mathcal{R}_{hf}$-radical algebras are just hypofinite algebras as well as $\mathcal{R}_{hc}$-radical algebras are hypocompact algebras. Now we note the following useful consequence of the heredity of radicals $\mathcal{R}_{hc}$ and $\mathcal{R}_{hf}$.

Corollary 3.60. *Every ideal of a hypocompact (respectively, hypofinite) normed algebra is a hypocompact (respectively, hypofinite) normed algebra.*

4. Multiplication operators on Banach bimodules

4.1. Banach bimodules

4.1.1. Elementary operators. Let A, B be Banach algebras and U a bimodule over A, B (shortly (A, B)-bimodule). Then in an obvious way U can be considered as an (A^1, B^1)-bimodule. We say that U is *a normed bimodule* if it is a normed space with a norm $\|\cdot\|_U$ and

$$\|aub\|_U \le \|a\|_A \|u\|_U \|b\|_B$$

for every $a \in A^1$, $b \in B^1$ and $u \in U$.

Let L_a and R_b be operators on U defined by $L_a x = ax$ and $R_b x = xb$ for every $x \in U$. By $\mathcal{E}\ell_{A,B}(U)$ we denote the algebra generated by all operators L_a, R_b. Its elements are called *elementary operators on U with coefficients in A, B*. If A, B are unital then $\mathcal{E}\ell_{A,B}(U)$ coincides with the algebra $\mathcal{E}_{A,B}(U)$ generated by all $L_a R_b$. In the general case $\mathcal{E}_{A,B}(U)$ is an ideal of $\mathcal{E}\ell_{A,B}(U)$ which is an ideal in $\mathcal{E}_{A^1,B^1}(U)$, and the latter can be regarded as a unitization of $\mathcal{E}\ell_{A,B}(U)$. Note also that

$$\mathcal{E}\ell_{A,B}(U) = \mathcal{E}_{A^1,B}(U) + \mathcal{E}_{A,B^1}(U).$$

Clearly operators in $\mathcal{E}_{A,B}(U)$ and $\mathcal{E}\ell_{A,B}(U)$ can be written in the form $T = \sum_{i=1}^n L_{a_i} R_{b_i}$ and, respectively, $T = L_a + R_b + \sum_{i=1}^n L_{a_i} R_{b_i}$, where $a, a_i \in A$, $b, b_i \in B_i$.

One may consider U as a left (B^{op})-module, where B^{op} is the algebra opposite to B. Then there is a natural homomorphism $\psi = \psi_U$ from $A \otimes_\gamma B^{\mathrm{op}}$ into the algebra $\mathcal{B}(U)$ of all continuous operators on U given by

$$\psi : z = \sum_{i=0}^n a_i \otimes b_i \longmapsto \sum_{i=0}^n L_{a_i} R_{b_i}$$

for every $a_i \in A$ and $b_i \in B$. Then

$$\|\psi(z) u\|_U \le \gamma(z) \|u\|_U \tag{4.1}$$

for every $u \in U$, whence $\ker \psi$ is closed. Since the image of ψ coincides with $\mathcal{E}_{A,B}(U)$, one may consider $\mathcal{E}_{A,B}(U)$ as a quotient of $A \otimes_\gamma B^{\mathrm{op}}$. This induces the quotient norm $\|\cdot\|_{\mathcal{E}_{A,B}}$, or simply $\|\cdot\|_{\mathcal{E}}$, on $\mathcal{E}_{A,B}(U)$ by

$$\|T\|_{\mathcal{E}} = \inf\left\{\sum \|a_i\|_A \|b_i\|_B : \sum L_{a_i} R_{b_i} = T\right\} \tag{4.2}$$

for every $T \in \mathcal{E}_{A,B}(U)$. So $\mathcal{E}_{A,B}(U)$ is a normed algebra with respect to $\|\cdot\|_{\mathcal{E}}$.

Proposition 4.1. *If U is a normed bimodule then $\|\cdot\|_{\mathcal{B}(U)} \le \|\cdot\|_{\mathcal{E}}$ on $\mathcal{E}_{A,B}(U)$, so $\mathcal{E}_{A,B}(U)$ is a normed subalgebra of $\mathcal{B}(U)$ with respect to $\|\cdot\|_{\mathcal{E}_{A,B}}$.*

Proof. Indeed, if $T \in \mathcal{E}_{A,B}(U)$ then $\|T\|_{\mathcal{B}(U)} \le \|T\|_{\mathcal{E}}$ by (4.1) and (4.2). □

In a similar way one can consider $\mathcal{E}_{A^1,B^1}(U)$ (and therefore $\mathcal{E}\ell_{A,B}(U)$) as a normed subalgebra of $\mathcal{B}(U)$. In the case of unital coefficient algebras A, B we don't distinguish $\mathcal{E}\ell_{A,B}(U)$ from $\mathcal{E}_{A,B}(U)$.

4.1.2. Multiplication operators. A normed bimodule is called *Banach* if it is a Banach space. It is clear that the completion $\widehat{U}$ of a normed bimodule U is a Banach bimodule and that one can identify $\mathcal{E}_{A,B}(U)$ and $\mathcal{E}_{A,B}\left(\widehat{U}\right)$.

Let U be a Banach (A, B)-bimodule. Let $\widehat{\mathcal{E}}_{A,B}(U)$ be the completion of $\mathcal{E}_{A,B}(U)$ in $\|\cdot\|_{\mathcal{E}}$. It is clear that $\widehat{\mathcal{E}}_{A,B}(U)$ is an algebra of continuous operators on U. The operators in $\widehat{\mathcal{E}}_{A,B}(U)$ are called *multiplication operators* on U.

Again, if U is a Banach bimodule then $\widehat{\mathcal{E}}_{A,B}(U) \subset \mathcal{B}(U)$ as usual, and $\widehat{\mathcal{E}}_{A,B}(U)$ is a Banach subalgebra of $\mathcal{B}(U)$.

Proposition 4.2. *If I and J are flexible ideals of A and B respectively, then $\mathcal{E}_{I,J}(U)$ is a flexible ideal of $\mathcal{E}_{A,B}(U)$, and $\widehat{\mathcal{E}}_{I,J}(U)$ with the norm $\|\cdot\|_{\mathcal{E}_{I,J}}$ is a Banach ideal of $\widehat{\mathcal{E}}_{A,B}(U)$.*

In what follows we often denote the coefficient algebras by A_1, A_2 instead of A, B.

Theorem 4.3. *Let U be a Banach bimodule over normed algebras A_1, A_2.*

(i) *If A_1 and A_2 are hypocompact then the algebra $\widehat{\mathcal{E}}_{A_1,A_2}(U)$ is hypocompact.*
(ii) *If at least one of the algebras A_i is tensor radical then $\widehat{\mathcal{E}}_{A_1,A_2}(U)$ is tensor radical.*
(iii) *If both A_i are commutative modulo the tensor radical then $\widehat{\mathcal{E}}_{A_1,A_2}(U)$ is commutative modulo the tensor radical.*

Proof. (i) Indeed, it is easy to see that $\widehat{\mathcal{E}}_{A_1,A_2}(U)$ is isometric to a quotient of $A_1\widehat{\otimes}A_2^{\mathrm{op}}$. So it is hypocompact by Theorem 3.57 and Corollary 3.47.

(ii) Since $\widehat{\mathcal{E}}_{A_1,A_2}(U)$ is isometric to the quotient of $A_1\widehat{\otimes}A_2^{\mathrm{op}}$ by the kernel of the natural map from $A_1\widehat{\otimes}A_2^{\mathrm{op}}$ into $\mathcal{B}(U)$, the statement follows from the fact that a quotient of a tensor radical normed algebra is tensor radical (which follows easily from Theorem 3.28).

(iii) Arguing as in (ii), we have only to prove that if a normed algebra B is commutative modulo the tensor radical then so is the quotient of B by a closed ideal J.

Let $q_J : B \longrightarrow B/J$ be the standard epimorphism. Then

$$q_J(\mathcal{R}_t(B)) \subset \mathcal{R}_t(B/J)$$

by Theorem 3.28. Assuming that $B/\mathcal{R}_t(B)$ is commutative, we obtain that

$$[a/J, b/J] = q_J([a,b]) \in \mathcal{R}_t(B/J) \qquad (a, b \in B).$$

This means that B/J is commutative modulo the tensor radical. □

Corollary 4.4. *Let U be a Banach bimodule over unital normed algebras A_1, A_2. If A_1 and A_2 are hypocompact then $\sigma_{\mathcal{B}(U)}(T)$ is (finite or) countable and*

$$\sigma_{\mathcal{B}(U)}(T) = \sigma_{\widehat{\mathcal{E}}_{A_1,A_2}(U)}(T)$$

for every $T \in \widehat{\mathcal{E}}_{A_1,A_2}(U)$.

Proof. Indeed, $\widehat{\mathcal{E}}_{A_1,A_2}(U)$ is a unital Banach subalgebra of $\mathcal{B}(U)$, and $\sigma_{\widehat{\mathcal{E}}_{A_1,A_2}(U)}(T)$ is countable (Theorem 3.50). So $\sigma_{\mathcal{B}(U)}(T) = \sigma_{\widehat{\mathcal{E}}_{A_1,A_2}(U)}(T)$ by Proposition 2.3. □

Now we consider the sum $\widehat{\mathcal{E}}_{A_1,I_2}(U) + \widehat{\mathcal{E}}_{I_1,A_2}(U)$ of Banach subalgebras of $\mathcal{B}(U)$ for closed ideals I_i of A_i.

Corollary 4.5. *Let I_i be a closed ideal of normed algebra A_i for $i = 1, 2$. If I_1 and I_2 are tensor radical then the algebra $\widehat{\mathcal{E}}_{A_1,I_2}(U) + \widehat{\mathcal{E}}_{I_1,A_2}(U)$ is tensor radical.*

Proof. Since $\widehat{\mathcal{E}}_{I_1,A_2}(U)$ and $\widehat{\mathcal{E}}_{A_1,I_2}(U)$ are ideals of $\widehat{\mathcal{E}}_{A_1,A_2}(U)$ and are tensor radical by Theorem 4.3, they are contained in $\mathcal{R}_t(\widehat{\mathcal{E}}_{A_1,A_2}(U))$. Since $\widehat{\mathcal{E}}_{A_1,I_2}(U)+\widehat{\mathcal{E}}_{I_1,A_2}(U)$ is a flexible ideal of $\widehat{\mathcal{E}}_{A_1,A_2}(U)$, it is also tensor radical by Proposition 3.40. □

Let us consider a more general situation.

Theorem 4.6. *Let $J_i \subset I_i$ be ideals of A_i, $i = 1,2$. Suppose that the algebras $\overline{I_i}/\overline{J_i}$ are tensor radical. Setting, for brevity, $\widehat{\mathcal{E}}_I = \widehat{\mathcal{E}}_{A_1,I_2}(U)+\widehat{\mathcal{E}}_{I_1,A_2}(U)$ and $\mathcal{E}_J = \mathcal{E}_{A_1,J_2}(U)+\mathcal{E}_{J_1,A_2}(U)$ we have that $\overline{\mathcal{E}_J}$ is an ideal of $\widehat{\mathcal{E}}_I$ and the algebra $\widehat{\mathcal{E}}_I/\overline{\mathcal{E}_J}$ is tensor radical.*

Proof. One can clearly assume that the ideals I_i and J_i are closed in A_i for $i = 1,2$.

Consider the Banach ideal $K_1 = (I_1/J_1)\,\widehat{\otimes}^{(\cdot)}\,(A_2/J_2)$ in the Banach algebra $B = (A_1/J_1)\,\widehat{\otimes}_\gamma\,(A_2/J_2)$. As K_1 is topologically isomorphic to a quotient of $(I_1/J_1)\,\widehat{\otimes}\,(A_2/J_2)$ then it is tensor radical. Similarly, the Banach ideal $K_2 = (A_1/J_1)\,\widehat{\otimes}^{(\cdot)}\,(I_2/J_2)$ in B is tensor radical. Then their flexible sum K_1+K_2 in B is tensor radical by Proposition 3.40.

Hence the quotient $(I_1\widehat{\otimes}^{(\cdot)}A_2+A_1\widehat{\otimes}^{(\cdot)}I_2)/\overline{J_1\otimes A_2+A_1\otimes J_2}$ is tensor radical because it is topologically isomorphic to K_1+K_2 by Proposition 2.11.

Consider now the natural epimorphism $\psi : I_1\widehat{\otimes}^{(\cdot)}A_2+A_1\widehat{\otimes}^{(\cdot)}I_2 \longrightarrow \widehat{\mathcal{E}}_I/\overline{\mathcal{E}_J}$. It is clear that $J_1\otimes A_2+A_1\otimes J_2 \subset \ker\psi$. So there is a bounded homomorphism from $(I_1\widehat{\otimes}^{(\cdot)}A_2+A_1\widehat{\otimes}^{(\cdot)}I_2)/\overline{J_1\otimes A_2+A_1\otimes J_2}$ onto $\widehat{\mathcal{E}}_I/\overline{\mathcal{E}_J}$. This epimorphism is open by the Banach Theorem. Therefore $\widehat{\mathcal{E}}_I/\overline{\mathcal{E}_J}$ is tensor radical by Theorem 3.28. □

The proved result is important for applications in Section 6. Note that Corollary 4.5 can be obtained from Theorem 4.6 if one takes $J_1 = J_2 = 0$.

In the following result we preserve notation of Theorem 4.6.

Corollary 4.7. *Let I and J be as in Theorem 4.6. Then $\widehat{\mathcal{E}}_I \subset Q_{\mathcal{E}_J}(\widehat{\mathcal{E}}_{A_1,A_2}(U))$. Thus if $T \in \widehat{\mathcal{E}}_I$ then, for every $\varepsilon > 0$, there are $m \in \mathbb{N}$ and elementary operators $S_n \in \mathcal{E}_J$ on U such that $\|T^n - S_n\|_{\mathcal{E}} < \varepsilon^n$ for every $n > m$.*

Proof. Follows from Proposition 2.1. □

4.2. Operator bimodules

Let us consider the case that A_1, A_2 are the algebras $\mathcal{B}(Y)$, $\mathcal{B}(X)$ of all bounded operators on Banach spaces X, Y. Let U be a normed subspace of $\mathcal{B}(X,Y)$ of all bounded operators from X to Y with the natural bounded $(\mathcal{B}(Y),\mathcal{B}(X))$-bimodule structure; we refer to it as a *normed subbimodule* of $\mathcal{B}(X,Y)$ or, simply, a *normed operator bimodule.* The latter means that U is supplied with its own norm $\|\cdot\|_U \geq \|\cdot\|_{\mathcal{B}} = \|\cdot\|$ and

$$\|axb\|_U \leq \|a\|\,\|x\|_U\,\|b\|$$

for all $a \in \mathcal{B}(Y)$, $b \in \mathcal{B}(X)$, $x \in U$. It is easy to see that if U is non-zero then U contains all finite rank operators. We also may assume that

$$\|x\|_U = \|x\|$$

for every rank one operator $x \in U$.

When U is complete in $\|\cdot\|_U$, one says that U is a *Banach operator bimodule*. In this case, for brevity, we also denote $\widehat{\mathcal{E}}_{\mathcal{B}(Y),\mathcal{B}(X)}(U)$ by $\widehat{\mathcal{B}}_*(U)$ and call its elements $(\mathcal{B})$-*multiplication operators* on U. It is clear that $\widehat{\mathcal{B}}_*(U)$ is a Banach subalgebra of $\mathcal{B}(U)$ with respect to the norm $\|\cdot\|_{\widehat{\mathcal{B}}_*} = \|\cdot\|_{\widehat{\mathcal{E}}_{\mathcal{B}(Y),\mathcal{B}(X)}}$.

Operator bimodules are closely related to operator ideals. If $\mathcal{U}$ is a Banach operator ideal in the sense of [20], e. g. the ideal $\mathcal{K}$ of compact operators or the ideal $\mathcal{N}$ of nuclear operators then each *component* $U = \mathcal{U}(X,Y)$ of $\mathcal{U}$ is a Banach operator bimodule. It can be proved that all Banach operator bimodules can be obtained in this way.

4.2.1. Semicompact multiplication and $(\mathcal{K})$-multiplication operators. The algebras A_i are semisimple so they have no radical ideals. But they can have pairs of ideals $J_i \subset I_i$ with radical quotients $I_i/\overline{J_i}$. Indeed, Lemma 3.54 shows that this is the case if we take the ideals $\mathcal{K}(X)$, $\mathcal{K}(Y)$ for I_i, and the ideals $\mathcal{F}(X)$, $\mathcal{F}(Y)$ for J_i. The possibility to use Theorem 4.6 is important for further applications. Before formulate the corresponding corollaries it will be convenient to introduce special terminology. Namely we set

$$\widehat{\mathcal{K}}_{\frac{1}{2}}(U) = \widehat{\mathcal{E}}_{\mathcal{B}(Y),\mathcal{K}(X)}(U) + \widehat{\mathcal{E}}_{\mathcal{K}(Y),\mathcal{B}(X)}(U),$$
$$\mathcal{F}_{\frac{1}{2}}(U) = \mathcal{E}_{\mathcal{B}(Y),\mathcal{F}(X)}(U) + \mathcal{E}_{\mathcal{F}(Y),\mathcal{B}(X)}(U).$$

Note that $\widehat{\mathcal{K}}_{\frac{1}{2}}(U)$ is taken here as the sum of Banach ideals $\widehat{\mathcal{E}}_{\mathcal{B}(Y),\mathcal{K}(X)}(U)$ and $\widehat{\mathcal{E}}_{\mathcal{K}(Y),\mathcal{B}(X)}(U)$ in $\widehat{\mathcal{B}}_*(U)$ with respective flexible norm (see Proposition 2.7). Thus $\widehat{\mathcal{K}}_{\frac{1}{2}}(U)$ consists of multiplication operators T on U of the form $\sum_{i=1}^{\infty} L_{a_i}R_{b_i}$ such that $\sum_i \|a_i\|\|b_i\| < \infty$ and at least one of the operators a_i or b_i is compact for every i. The norm $\|T\|_{\widehat{\mathcal{K}}_{\frac{1}{2}}}$ is equal to $\inf \sum_i \|a_i\|\|b_i\|$ for all such representations of T. We call operators in $\widehat{\mathcal{K}}_{\frac{1}{2}}(U)$ *semicompact multiplication operators.*

Similarly, operators in $\mathcal{F}_{\frac{1}{2}}(U)$ are called *semifinite elementary operators.* They are just the elementary operators $\sum_{i=1}^{n} L_{a_i}R_{b_i}$ where a_i or b_i is a finite rank operator for each i.

As a concrete application of Theorem 4.6, we obtain the following

Corollary 4.8. *Let $U \subset \mathcal{B}(X,Y)$ be a Banach operator bimodule. Then the algebra $\widehat{\mathcal{K}}_{\frac{1}{2}}(U)/\overline{\mathcal{F}_{\frac{1}{2}}(U)}$ is tensor radical.*

As a consequence we have the following

Corollary 4.9. *Let $U \subset \mathcal{B}(X,Y)$ be a Banach operator bimodule. Then*

$$\widehat{\mathcal{K}}_{\frac{1}{2}}(U) \subset Q_{\mathcal{F}_{\frac{1}{2}}(U)}(\widehat{\mathcal{B}}_*(U)).$$

Since the norm in $\widehat{\mathcal{B}}_*(U)$ majorizes the operator norm in $\mathcal{B}(U)$, we also obtain the following result.

Corollary 4.10. *Let T be a semicompact multiplication operator on a Banach operator bimodule U. Then for each $\varepsilon > 0$ there are $m \in \mathbb{N}$ and semifinite elementary operators S_n on U such that $\|T^n - S_n\|_{\mathcal{B}(U)} < \varepsilon^n$ for every $n > m$.*

The other useful ideal in $\widehat{\mathcal{B}}_*(U)$, namely the ideal $\widehat{\mathcal{K}}_*(U)$ of $(\mathcal{K})$-*multiplication operators,* is defined by

$$\widehat{\mathcal{K}}_*(U) = \widehat{\mathcal{E}}_{\mathcal{K}(Y),\mathcal{K}(X)}(U).$$

In many cases, for instance when the norm of U coincides with the operator one, $\widehat{\mathcal{K}}_*(U)$ consists of compact operators on U. It is clear that $\widehat{\mathcal{K}}_*(U)$ is a Banach ideal of $\widehat{\mathcal{K}}_{\frac{1}{2}}(U)$ and $\widehat{\mathcal{B}}_*(U)$ with respect to the norm $\|\cdot\|_{\widehat{\mathcal{K}}_*} = \|\cdot\|_{\widehat{\mathcal{E}}_{\mathcal{K}(Y),\mathcal{K}(X)}}$.

Proposition 4.11. *$\widehat{\mathcal{K}}_*(U)$ is a bicompact Banach algebra for every Banach operator bimodule U.*

Proof. Indeed, $\widehat{\mathcal{K}}_*(U)$ is topologically isomorphic to a quotient of the projective tensor product of bicompact algebras $\mathcal{K}(Y)$ and $\mathcal{K}(X)^{\mathrm{op}}$ which is bicompact itself by Lemma 3.56. □

Corollary 4.12. *Let U be a Banach operator bimodule. Then $\sigma_{\mathcal{B}(U)}(T)$ is at most countable and $\sigma_{\mathcal{B}(U)}(T) \cup \{0\} = \sigma_{\widehat{\mathcal{K}}_*(U)}(T) \cup \{0\} = \sigma_{\widehat{\mathcal{B}}_*(U)}(T) \cup \{0\}$ for every $T \in \widehat{\mathcal{K}}_*(U)$.*

Proof. Indeed, $\widehat{\mathcal{K}}_*(U)$ is a Banach subalgebra of $\mathcal{B}(U)$ and is a Banach ideal of $\widehat{\mathcal{B}}_*(U)$. Since $\sigma_{\widehat{\mathcal{K}}_*(U)}(T)$ is countable by [1, Theorem 4.4], then it is easy to see that $\sigma_{\mathcal{B}(U)}(T) \cup \{0\} = \sigma_{\widehat{\mathcal{K}}_*(U)}(T) \cup \{0\}$ by Proposition 2.3. Further, we have that $\sigma_{\widehat{\mathcal{K}}_*(U)}(T) \cup \{0\} = \sigma_{\widehat{\mathcal{B}}_*(U)}(T) \cup \{0\}$ by Remark 2.4. □

4.2.2. $(\mathcal{N})$-multiplication operators and trace. The norms and spectra of elementary operators were studied in many works. Here we would like to mention a simple formula for their traces.

Let $\mathcal{N}$ be the operator ideal of nuclear operators. Recall that every operator $a \in \mathcal{N}(X,Y)$ has a representation $\sum f_i \otimes x_i$ with $\sum \|f_i\| \|x_i\| < \infty$ for $f_i \in X^*$ and $x_i \in Y$. The nuclear norm $\|\cdot\|_{\mathcal{N}}$ is given by

$$\|a\|_{\mathcal{N}} = \inf\left\{\sum \|f_i\| \|x_i\| : \sum f_i \otimes x_i = a\right\}.$$

So, as is well known, the projective tensor product $X^* \widehat{\otimes}_\gamma Y$ is identified with $\mathcal{N}(X,Y)$. If $X = Y$, then the trace of a is defined by

$$\operatorname{trace}(a) = \sum f_i(x_i).$$

Let $U \subset \mathcal{B}(X,Y)$ be a Banach operator bimodule. One can define the ideal $\mathcal{N}_*(U)$ of *($\mathcal{N}$)-multiplication operators* by

$$\widehat{\mathcal{N}}_*(U) = \widehat{\mathcal{E}}_{\mathcal{N}(Y),\mathcal{N}(X)}(U)$$

with respect to the norm $\|\cdot\|_{\widehat{\mathcal{N}}_*} = \|\cdot\|_{\widehat{\mathcal{E}}_{\mathcal{N}(Y),\mathcal{N}(X)}}$.

Proposition 4.13. *Every ($\mathcal{N}$)-multiplication operator T on U is nuclear. If $T = \sum L_{a_i} R_{b_i}$ with $\sum \|a_i\|_{\mathcal{N}} \|b_i\|_{\mathcal{N}} < \infty$ for $a_i, b_i \in \mathcal{N}$, then*

$$\operatorname{trace}(T) = \sum_i \operatorname{trace}(a_i)\operatorname{trace}(b_i).$$

Proof. Assume first that $a = f \otimes x$ and $b = g \otimes y$ are rank one operators. For every $u \in U$, we obtain that

$$(f \otimes x)\, u\, (g \otimes y) = f(uy)\, g \otimes x.$$

It is clear that the map $u \longmapsto f(uy)$ is a bounded linear functional on U. Indeed,

$$|f(uy)| \leq \|f\| \|u\| \|y\| \leq (\|f\| \|y\|) \|u\|_U .$$

In the same time, $g \otimes x \in U$ for every $g \in X^*$ and $x \in Y$. So $L_a R_b$ is a rank one operator.

This also shows that if a_i and b_i are nuclear then $L_{a_i} R_{b_i}$ is nuclear, because an absolutely convergent series of nuclear operators is nuclear. Therefore T is nuclear by the same reason.

Clearly

$$\operatorname{trace}(L_{f\otimes x} R_{g\otimes y}) = f((g \otimes x)\, y) = f(x)\, g(y) = \operatorname{trace}(f \otimes x)\operatorname{trace}(g \otimes y).$$

Therefore, for nuclear $a_i = \sum f_j \otimes x_j$ and $b_i = \sum g_k \otimes y_k$, we obtain that

$$\begin{aligned}\operatorname{trace}(L_{a_i} R_{b_i}) &= \sum_{j,k} \operatorname{trace}(L_{f_j\otimes x_j} R_{g_k \otimes y_k}) = \sum_{j,k} \operatorname{trace}(f_j \otimes x_j)\operatorname{trace}(g_k \otimes y_k)\\ &= \operatorname{trace}(a_i)\operatorname{trace}(b_i),\end{aligned}$$

whence

$$\operatorname{trace}(T) = \sum \operatorname{trace}(L_{a_i} R_{b_i}) = \sum \operatorname{trace}(a_i)\operatorname{trace}(b_i). \qquad \square$$

4.3. Some constructions related to multiplication operators

4.3.1. Integral operators. Let $U \subset \mathcal{B}(X,Y)$ be a Banach operator bimodule. Assume first that U is reflexive as a Banach space. Let (Ω, μ) be a measure space, and let $L_2^{\mathcal{B}(Y)}(\Omega,\mu)$ be the space of all measurable $\mathcal{B}(Y)$-valued functions $a : \omega \longmapsto a(\omega)$ with the norm

$$\|a\|_{L_2} = \int_\Omega \|a(\omega)\|^2 \, d\mu < \infty.$$

For any two functions $\omega \longmapsto a(\omega) \in L_2^{\mathcal{B}(Y)}(\Omega,\mu)$ and $\omega \longmapsto b(\omega) \in L_2^{\mathcal{B}(X)}(\Omega,\mu)$, one can define an operator $T_{a,b}$ on U by means of the Bochner integral [15, Section

3.3.7]

$$T_{a,b}(x) = \int_{\Omega} a(\omega)xb(\omega)d\mu.$$

Note that one may define $L_2^{\mathcal{K}(Y)}(\Omega,\mu)$ if replace $\mathcal{B}(Y)$ by $\mathcal{K}(Y)$, and consider $L_2^{\mathcal{K}(Y)}(\Omega,\mu)$ as a subspace of $L_2^{\mathcal{B}(Y)}(\Omega,\mu)$.

Proposition 4.14. *Let $U \subset \mathcal{B}(X,Y)$ be a Banach operator bimodule, and let U be reflexive as a Banach space. Then $T_{a,b} \in \widehat{\mathcal{E}}_{\mathcal{B}(Y),\mathcal{B}(X)}$ and $\|T_{a,b}\|_{\widehat{\mathcal{E}}_{\mathcal{B}(Y),\mathcal{B}(X)}} \le \|a\|_{L_2}\|b\|_{L_2}$. If in particular $\omega \longmapsto a(\omega) \in L_2^{\mathcal{K}(Y)}(\Omega,\mu)$ (respectively, $\omega \longmapsto b(\omega) \in L_2^{\mathcal{K}(X)}(\Omega,\mu)$) then $T_{a,b} \in \widehat{\mathcal{E}}_{\mathcal{K}(Y),\mathcal{B}(X)}$ and $\|T_{a,b}\|_{\widehat{\mathcal{E}}_{\mathcal{K}(Y),\mathcal{B}(X)}} \le \|a\|_{L_2}\|b\|_{L_2}$ (respectively, $T_{a,b} \in \widehat{\mathcal{E}}_{\mathcal{B}(Y),\mathcal{K}(X)}$ and $\|T_{a,b}\|_{\widehat{\mathcal{E}}_{\mathcal{B}(Y),\mathcal{K}(X)}} \le \|a\|_{L_2}\|b\|_{L_2}$).*

Proof. Let $a_n(\omega) \in L_2^{\mathcal{B}(Y)}(\Omega,\mu)$ be a sequence of simple functions that tend to $a(\omega)$ in norm of $L_2^{\mathcal{B}(Y)}(\Omega,\mu)$. Clearly $T_{a,b}$ is the limit of operators $T_{a_n,b}$ in the norm topology of $\mathcal{B}(U)$.

Each a_n is a finite sum of functions $k_i\varphi_i(t)$, where $k_i \in \mathcal{B}(Y)$ and φ_i is the characteristic function of a measurable set $\Lambda_i \subset \Omega$. Hence

$$T_{a_n,b} = \sum_i L_{k_i}R_{t_i}$$

where $t_i = \int_{\Lambda_i} b(t)d\mu \in \mathcal{B}(X)$. Therefore $T_{a_n,b} \in \mathcal{E}_{\mathcal{B}(Y),\mathcal{B}(X)}$ and

$$\|T_{a_n,b}\|_{\widehat{\mathcal{E}}_{\mathcal{B}(Y),\mathcal{B}(X)}} \le \|a_n\|_{L_2}\|b\|_{L_2}.$$

It follows from this that the sequence of operators $T_{a_n,b}$ is fundamental in $\widehat{\mathcal{E}}_{\mathcal{B}(Y),\mathcal{B}(X)}$, so it tends to some element $T \in \widehat{\mathcal{E}}_{\mathcal{B}(Y),\mathcal{B}(X)}$, and $\|T\|_{\widehat{\mathcal{E}}_{\mathcal{B}(Y),\mathcal{B}(X)}} \le \|a\|_{L_2}\|b\|_{L_2}$. By the above, $T = T_{a,b}$ and we are done.

The other statements are proved similarly. □

Let $a \in L_2^{\mathcal{K}(Y)}(\Omega,\mu)$, $b \in L_2^{\mathcal{K}(X)}(\Omega,\mu)$, $s \in L_2^{\mathcal{B}(Y)}(\Omega,\mu)$, $t \in L_2^{\mathcal{B}(X)}(\Omega,\mu)$. Then the operator $T_{a,b,s,t}$ defined on a Banach operator bimodule $U \subset \mathcal{B}(X,Y)$ by the formula

$$T_{a,b,s,t}(x) = \int_I a(\omega)xt(\omega)d\mu + \int_I s(\omega)xb(\omega)d\mu,$$

is called an *integral semicompact operator*. Indeed, it follows from Proposition 4.14 that this operator is semicompact multiplication operator.

If we wish to remove the restriction of reflexivity of U and still have that $T_{a,b,s,t}$ belongs to $\widehat{\mathcal{K}}_{\frac{1}{2}}(U)$, we should impose continuity conditions which allow us to deal with Riemann integral sums (see for instance [15, Section 3.3.7]). For brevity we will formulate the corresponding result in a form which is far from the most general.

Theorem 4.15. *Let $U \subset \mathcal{B}(X,Y)$ be a Banach operator bimodule, and let $(\Omega,\mu) = (I,\mu)$, where μ is a regular measure on an interval $I \subset \mathbb{R}$. Then every integral semicompact operator $T_{a,b,s,t}$ belongs to $\widehat{\mathcal{K}}_{\frac{1}{2}}(U)$ and*

$$\|T_{a,b,s,t}\|_{\widehat{\mathcal{K}}_{\frac{1}{2}}(U)} \leq \|a\|_{L_2}\|t\|_{L_2} + \|b\|_{L_2}\|s\|_{L_2}.$$

4.3.2. Matrix multiplication operators. Let $(T_{ji})_{j,i=1}^n$ be a matrix of multiplication operators on a Banach operator bimodule $U \subset \mathcal{B}(X,Y)$. It defines an operator $T = [T_{ij}]$ on $U^{(n)}$ by the formula

$$T(x_1,\dots,x_n) = (y_1,\dots,y_n) \text{ where } y_i = \sum_j T_{ij}x_j.$$

Let us denote the algebra of all such operators by $\mathbb{M}_n(\widehat{\mathcal{E}}_{\mathcal{B}(X),\mathcal{B}(Y)}(U))$. Also, by $\mathbb{M}_n(\widehat{\mathcal{K}}_{\frac{1}{2}}(U))$ we denote the ideal of $\mathbb{M}_n(\widehat{\mathcal{E}}_{\mathcal{B}(X),\mathcal{B}(Y)}(U))$ which consists of all operators $T = [T_{ij}]$ with $T_{ij} \in \widehat{\mathcal{K}}_{\frac{1}{2}}(U)$ for $1 \leq i,j \leq n$. In a similar way we define the subspace $\mathbb{M}_n(\mathcal{F}_{\frac{1}{2}}(U))$ of $\mathbb{M}_n(\widehat{\mathcal{K}}_{\frac{1}{2}}(U))$. The closure $\overline{\mathbb{M}_n(\mathcal{F}_{\frac{1}{2}}(U))}$ of $\mathbb{M}_n(\mathcal{F}_{\frac{1}{2}}(U))$ in $\mathbb{M}_n(\widehat{\mathcal{K}}_{\frac{1}{2}}(U))$ is an ideal of $\mathbb{M}_n(\widehat{\mathcal{K}}_{\frac{1}{2}}(U))$.

Theorem 4.16. *Let $U \subset \mathcal{B}(X,Y)$ be a Banach operator bimodule. Then the algebra $\mathbb{M}_n(\widehat{\mathcal{K}}_{\frac{1}{2}}(U))/\overline{\mathbb{M}_n(\mathcal{F}_{\frac{1}{2}}(U))}$ is tensor radical.*

Proof. The algebra $\mathbb{M}_n(\widehat{\mathcal{K}}_{\frac{1}{2}}(U))$ is topologically isomorphic to $\mathbb{M}_n \otimes_\gamma (\widehat{\mathcal{K}}_{\frac{1}{2}}(U))$, where $\mathbb{M}_n$ is the algebra of $n \times n$ matrices. Furthermore, the algebra $\overline{\mathbb{M}_n(\mathcal{F}_{\frac{1}{2}}(U))}$ is topologically isomorphic to $\mathbb{M}_n \otimes_\gamma \overline{\mathcal{F}_{\frac{1}{2}}(U)}$. Hence, for every Banach algebra A, we have that

$$\begin{aligned}
\left(\mathbb{M}_n\left(\widehat{\mathcal{K}}_{\frac{1}{2}}(U)\right)/\overline{\mathbb{M}_n\left(\mathcal{F}_{\frac{1}{2}}(U)\right)}\right)\widehat{\otimes}A &\cong \left(\left(\mathbb{M}_n \otimes_\gamma \widehat{\mathcal{K}}_{\frac{1}{2}}(U)\right)/\left(\mathbb{M}_n \otimes_\gamma \overline{\mathcal{F}_{\frac{1}{2}}(U)}\right)\right)\widehat{\otimes}A \\
&\cong \left(\mathbb{M}_n \otimes_\gamma \left(\widehat{\mathcal{K}}_{\frac{1}{2}}(U)/\overline{\mathcal{F}_{\frac{1}{2}}(U)}\right)\right)\widehat{\otimes}A \\
&\cong \left(\widehat{\mathcal{K}}_{\frac{1}{2}}(U)/\overline{\mathcal{F}_{\frac{1}{2}}(U)}\right)\widehat{\otimes}\left(\mathbb{M}_n \otimes_\gamma A\right)
\end{aligned}$$

The latter algebra is radical because $\widehat{\mathcal{K}}_{\frac{1}{2}}(U)/\overline{\mathcal{F}_{\frac{1}{2}}(U)}$ is tensor radical by Corollary 4.8. □

5. Multiplication operators on algebras satisfying compactness conditions

In this section we consider elementary and multiplication operators in the most popular meaning: as elementary and multiplication operators with coefficients in a Banach algebra acting on the algebra itself. In terms of the previous section, we consider the case $A_1 = A_2 = U = A$, that is we regard A as an A-bimodule. For brevity we remove the indication of a bimodule in our standard notation for the multiplication algebra: we write $\widehat{\mathcal{E}}_{A,A}$ instead of $\widehat{\mathcal{E}}_{A,A}(A)$ (taking the occasion to use $\widehat{\mathcal{E}}_{I,J}$ for ideals $I, J \subset A$). Furthermore, we denote the algebra of all elementary operators on A by $\mathcal{E}\ell(A)$ instead of $\mathcal{E}\ell_{A,A}(A)$.

To make our assumptions more concrete we impose various compactness conditions on A. As we know, even the weakest of them, the hypocompactness of A, implies that $\mathrm{Rad}(A)$ coincides with $\mathcal{R}_t(A)$.

5.1. Multiplication operators on algebras commutative modulo the radical

Since radical hypocompact Banach algebras are tensor radical, we can apply results of Section 4.

Corollary 5.1. *If A is a hypocompact Banach algebra then the algebra $\widehat{\mathcal{E}}_{\mathrm{Rad}(A),A} + \widehat{\mathcal{E}}_{A,\mathrm{Rad}(A)}$ is tensor radical and hypocompact.*

Proof. The first statement follows from Corollary 4.5. Arguing as in the proof of Theorem 4.3 with using Theorem 3.57 and Corollary 3.47, we prove the second statement. □

In particular all elementary operators $L_a + R_b + \sum_{i=1}^n L_{a_i} R_{b_i}$ with a_i or b_i in $\mathrm{Rad}(A)$ for each i are quasinilpotent elements of $\widehat{\mathcal{E}}_{A,A}$. Since the norm in $\widehat{\mathcal{E}}_{A,A}$ majorizes the operator norm, they are quasinilpotent operators on A.

Below by spectra of elementary operators we mean their spectra in the algebra $\mathcal{B}(A)$ of all bounded operators on A. Clearly the unitization $\mathcal{E}\ell(A)^1$ of $\mathcal{E}\ell(A)$ consists of elements of the form $\sum_{i=1}^n L_{a_i} R_{b_i}$ where $a_i, b_i \in A^1$.

Theorem 5.2. *Let A be a hypocompact Banach algebra. If $u = \sum_{i=1}^n L_{a_i} R_{b_i} \in \mathcal{E}\ell(A)^1$, $v = \sum_{j=1}^m L_{c_j} R_{b_j} \in \mathcal{E}\ell(A)^1$ and all commutators $[a_i, c_j]$ and $[b_i, d_j]$ belong to $\mathrm{Rad}(A)$ then*

$$\sigma(u+v) \subset \sigma(u) + \sigma(v) \tag{5.1}$$

and

$$\sigma(uv) \subset \sigma(u)\sigma(v). \tag{5.2}$$

Proof. Let us denote A^1 by B and $\mathrm{Rad}(A)$ by J for brevity. Let $C = B \widehat{\otimes} B^{\mathrm{op}}$ and $E = J \otimes B^{\mathrm{op}} + B \otimes J^{\mathrm{op}}$. As $J = \mathcal{R}_t(A)$ by Theorem 3.52, we have that $E \subset \mathrm{Rad}(C)$. Setting $u' = \sum_i a_i \otimes b_i$ and $v' = \sum_j c_j \otimes b_j$, we have that $[u', v'] \in E$ (because $[a \otimes b, c \otimes d] = [a, c] \otimes bd + ca \otimes [b, d]$). So these elements commute modulo the radical of C.

Let $\phi : C \longrightarrow \mathcal{B}(A)$ be the homomorphism sending $a \otimes b$ to $L_a R_b$ for every $a, b \in B$. Then the algebra $D = \phi(C)$ supplied with the norm of the quotient $C / \ker \phi$ is a Banach subalgebra of $\mathcal{B}(A)$. The elements $u = \phi(u')$ and $v = \phi(v')$ commute modulo $\phi(\mathrm{Rad}(C)) \subset \mathrm{Rad}(D)$. Hence

$$\sigma_D(u+v) \subset \sigma_D(u) + \sigma_D(v) \tag{5.3}$$

and

$$\sigma_D(uv) \subset \sigma_D(u)\sigma_D(v), \tag{5.4}$$

where $\sigma_D(x)$ denotes the spectrum of $x \in D$ with respect to D.

As C is hypocompact by Theorem 3.57, and D is isomorphic to a quotient of C, then D is hypocompact by Corollary 3.47. By Theorem 3.50, $\sigma_D(u)$ and $\sigma_D(v)$ are finite or countable. Using Proposition 2.3, we get that $\sigma_D(u) = \sigma(u)$ and $\sigma_D(v) = \sigma(v)$. So the inclusions (5.3) and (5.4) imply (5.1) and (5.2). $\square$

Let A be a Banach algebra. Recall that the center modulo the radical or "Rad-center" $Z_{\mathrm{Rad}}(A)$ is the set $\{a \in A : [a,x] \in \mathrm{Rad}(A) \text{ for all } x \in A\}$.

Corollary 5.3. *If A is a hypocompact Banach algebra and $u \in L_{Z_{\mathrm{Rad}}(A)}R_A + L_A R_{Z_{\mathrm{Rad}}(A)}$ then inclusions* (5.1) *and* (5.2) *hold for all $v \in \mathcal{E\ell}(A)$.*

Let us call a subalgebra B of a Banach algebra *spectrally computable* if inclusions (5.1) and (5.2) hold for all elements $u, v \in B$.

Corollary 5.4. *If A is a hypocompact Banach algebra commutative modulo the radical then $\mathcal{E\ell}(A)$ is a spectrally computable subalgebra of $\mathcal{B}(A)$.*

Remark 5.5. *In virtue of Corollary* 4.4 *and by continuity of the spectrum on operators with countable spectra, the previous results as well as the results of Section* 5.2 *can be extended to multiplication operators. But we prefer to present them in less general and more traditional setting of elementary operators.*

5.2. Engel algebras

A Banach Lie algebra $\mathcal{L}$ is called *Engel* if all operators $\mathrm{ad}_{\mathcal{L}}(a) : x \to [a,x]$ on $\mathcal{L}$ are quasinilpotent. This is a natural functional-analytic extension of the class of nilpotent Lie algebras because the latter can be defined as Lie algebras for which all operators $\mathrm{ad}_{\mathcal{L}}(a)$ are nilpotent of some restricted order [38]. A Banach algebra A is said to be *Engel* if it is Engel as a Banach Lie algebra that is if all operators $L_a - R_a$ are quasinilpotent. It is proved in [28, Proposition 5.21] (and can be easily deduced from a more general result of Aupetit and Mathieu [3]) that all Engel Banach algebras are commutative modulo the radical. We call A *strongly Engel* if

$$\sigma\left(\sum_{i=1}^{n} L_{a_i} R_{b_i}\right) \subset \sigma\left(\sum_{i=1}^{n} a_i b_i\right)$$

for all $a_i, b_i \in A^1$ and $n \in \mathbb{N}$. It will be shown below that for hypocompact Banach algebras these notions coincide.

Theorem 5.6. *Let A be a hypocompact Banach algebra commutative modulo the radical. Suppose that A is generated (as a Banach algebra) by a subset M such that the operators $L_a - R_a$ are quasinilpotent for all $a \in M$. Then A is strongly Engel.*

Proof. By Corollary 5.4, the algebra $\mathcal{E\ell}(A)$ is spectrally computable. Using this fact, it can be easily shown that the set E of all $a \in A$, for which $\sigma(L_a - R_a) = \{0\}$ is a subalgebra of A. Indeed, if $a, b \in E$ then

$$\begin{aligned}\sigma(L_{a+b} - R_{a+b}) &= \sigma(L_a - R_a + L_b - R_b) \\ &\subset \sigma(L_a - R_a) + \sigma(L_b - R_b) = \{0\},\end{aligned}$$

so $a+b \in E$. Furthermore,

$$\begin{aligned}\sigma(L_{ab}-R_{ab}) &= \sigma(L_a(L_b-R_b)+(L_a-R_a)R_b+R_{ab-ba}) \\ &\subset \sigma(L_a)\sigma(L_b-R_b))+\sigma(L_a-R_a)\sigma(R_b)+\sigma(R_{ab-ba})=\{0\}\end{aligned}$$

whence $ab \in E$.

Since A is generated by M and $M \subset E$, the subalgebra E is dense in A.

Since A is hypocompact, its elements have countable spectra by Theorem 3.50 and therefore the spectra of all operators $L_a - R_a$ are countable. Hence they are the points of continuity of the spectral radius. It follows that E is closed whence $E = A$. We proved that A is an Engel algebra.

To see that A is strongly Engel, note that an operator $\sum_i^n L_{a_i}R_{b_i}$ can be written as $\sum_i^n L_{a_i}(R_{b_i}-L_{b_i})+L_c$, where $c=\sum_i^n a_ib_i$. Since $\mathcal{E}\ell(A)$ is spectrally computable and $\sigma(R_{b_i}-L_{b_i})=\{0\}$, we obtain that

$$\sigma\Big(\sum_i^n L_{a_i}R_{b_i}\Big)\subset\sigma(L_c)\subset\sigma(c). \qquad\square$$

Corollary 5.7. *Let A be a hypocompact Banach algebra generated by an Engel closed Lie subalgebra $\mathcal{L}$. Then A is a strongly Engel Banach algebra commutative modulo the radical.*

Proof. Without loss of generality, we consider the case when A is unital and generated as a Banach algebra by $\mathcal{L}$ and the identity element 1. Let us show first that all operators $L_a - R_a$ with $a \in \mathcal{L}$ are quasinilpotent on A (by our assumptions, they are quasinilpotent on $\mathcal{L}$). Indeed, they are bounded derivations of A and their spectra are countable (because spectra of elements of A are countable by Theorem 3.50). By [28, Corollary 3.7], they are quasinilpotent on the closed subalgebra of A generated by $\mathcal{L}$, that is on A. Taking into account Theorem 5.6, we have only to prove that A is commutative modulo the radical.

Let π be a strictly irreducible representation of A on X. We will obtain a contradiction assuming that $\dim(X) > 1$. Changing A by $A/\ker\pi$, one may suppose that π is faithful.

We already know that spectra of elements of A are countable. Now we claim that the spectra of elements of $\mathcal{L}$ are one-point. Indeed, if $\sigma(a)$ is not a singleton for some element $a \in \mathcal{L}$ then it is not connected and, by [28, Proposition 3.16], a has a non-trivial Riesz projection p commuting with $\mathcal{L}$. Hence p is in the center of A, which is impossible, because A is a primitive Banach algebra.

Define the function $h : \mathcal{L} \longrightarrow \mathbb{C}$ by $h(a) = \lambda$ if $\sigma(a) = \{\lambda\}$, for every $a \in \mathcal{L}$. We claim that h is a character of $\mathcal{L}$, i.e., h is a bounded linear functional on $\mathcal{L}$ that vanishes on $[\mathcal{L},\mathcal{L}]$. Indeed, by using Proposition 3.48, one can find a proper closed ideal J of A such that A/J is a compact Banach algebra. As $1/J$ is a compact element, then clearly it is a finite rank element. Hence A/J is finite dimensional. As $\sigma(a/J) \subset \sigma(a)$ for every $a \in \mathcal{L}$, then one can define the function $g : \mathcal{L}/J \longrightarrow \mathbb{C}$ by $g(a/J) = \lambda$ if $\sigma(a/J) = \{\lambda\}$ for every $a \in \mathcal{L}$. It is clear that $g(a/J) = h(a)$ for every $a \in \mathcal{L}$ and $\mathcal{L}/J$ is a nilpotent Lie subalgebra of A/J, whence A/J is commutative modulo the radical by the Engel theorem and g is a character of $\mathcal{L}/J$. So h is a character of $\mathcal{L}$.

Now, replacing every element a of $\mathcal{L}$ with $\sigma(a) = \{\lambda\}$ by $a - \lambda$, one may assume that $\mathcal{L}$ consists of quasinilpotent elements. Then

$$G = \exp(\mathcal{L}) = \{\exp(a) : a \in \mathcal{L}\}$$

is a group by [37], and $\sigma(b) = \{1\}$ for each $b \in G$. It follows that $r(M) = 1$ for every precompact subset M of G. By Corollary 3.51, $\rho(M) = 1$. In particular,

$$\rho(M \cup \{1, \exp(\lambda a)\}) = 1 \tag{5.5}$$

for every $a \in \mathcal{L}$ and $\lambda \in \mathbb{C}$.

Choose an arbitrary element $a \in \mathcal{L}$ and define the function f on $\mathbb{C}$ by $f(\lambda) = \rho(M(\exp(\lambda a) - 1)/\lambda)$ for a fixed precompact subset M of G. This function is subharmonic by [25, Theorem 3.5] and tends to zero when $\lambda \to \infty$, because

$$\begin{aligned}
\rho\left(M\left(\exp(\lambda a) - 1\right)\right) &\leq \rho((M \cup \{1\})\exp(\lambda a) - M \cup \{1\}) \\
&\leq \rho((M \cup \{1, \exp(\lambda a)\})^2 - (M \cup \{1, \exp(\lambda a)\})^2) \\
&\leq \rho\left(2\,\mathrm{abs}\left((M \cup \{1, \exp(\lambda a)\})^2\right)\right) \\
&= 2\rho\left((M \cup \{1, \exp(\lambda a)\})^2\right) \\
&= 2\rho\left(M \cup \{1, \exp(\lambda a)\}\right)^2 = 2,
\end{aligned}$$

by (5.5), where $\mathrm{abs}\,(S)$ denotes the absolutely convex hull of a bounded set $S \subset A$ and the equality $\rho(\mathrm{abs}\,(S)) = \rho(S)$ [25, Proposition 2.6] easily follows from the characterization of $\rho(S)$ as infimum of $\|S\|'$ when $\|\cdot\|'$ runs over all algebra norms equivalent to given one. Hence f is constant and, moreover, $f(\lambda) = 0$ for all λ. In particular, $f(0) = 0$, whence it is easy to see that

$$\rho(aM) = 0.$$

Now if an element x belongs to the linear span of M then

$$\rho(ax) = 0$$

by Lemma 3.44. Since linear span $\mathrm{lin}(G)$ of G is a subalgebra of A and the closure of $\mathrm{lin}(G)$ contains $\mathcal{L}$, we conclude that $\mathrm{lin}(G)$ is dense in A. The previous argument shows that

$$\rho(ax) = 0$$

for all $x \in \mathrm{lin}(G)$. Since spectra of elements of A are countable, the spectral radius is continuous on A, whence

$$\rho(ax) = 0$$

for all $x \in A$. This means that $a \in \mathrm{Rad}(A)$. So $\mathcal{L} \subset \mathrm{Rad}(A)$ and the closed algebra generated by $\mathcal{L}$ is radical. Then $\dim(X) = 1$, a contradiction. □

5.3. Generalized multiplication operators

It follows from the above that if A is a radical hypocompact algebra then all operators in $\mathcal{E\ell}(A)$ and, more generally, in $\widehat{\mathcal{E}}_{A,A}$ are quasinilpotent. Here we discuss the possibility to extend this result to operators in the norm-closure $\mathrm{Mul}(A)$ of $\mathcal{E\ell}(A)$ in $\mathcal{B}(A)$. Note first of all that it is an open problem if $\mathrm{Mul}(A)$ is radical even for bicompact A (and even for the case that A is a radical closed algebra of compact operators). We call elements of $\mathrm{Mul}(A)$ *generalized multiplication operators.*

Theorem 5.8. *If A is a compact algebra and $J = \mathrm{Rad}(A)$ then the closed ideal I generated by $L_J R_A \cup L_A R_J$ is contained in $\mathrm{Rad}(\mathrm{Mul}(A))$. As a consequence, $L_J + R_J + I$ consists of quasinilpotents.*

Proof. Since A is compact, $L_{a^2} R_b$ is compact for any $a, b \in A$. Indeed, since

$$L_a R_b + L_b R_a = 1/2(L_{a+b} R_{a+b} - L_{a-b} R_{a-b})$$

is a compact operator, the same is true for

$$L_{a^2} R_b = L_a(L_a R_b + L_b R_a) - (L_a R_a) L_b.$$

Now it follows from Corollary 5.1 that if $b \in J$ then $L_{a^2} R_b T$ is a compact quasinilpotent operator for every elementary operator T on A. By continuity of the spectral radius, the same is true for all $T \in \mathrm{Mul}(A)$. Hence $L_{a^2} R_b \in \mathrm{Rad}(\mathrm{Mul}(A))$.

By the Nagata–Higman theorem (for $n = 2$), every product $a_1 a_2 a_3$ of elements of A can be represented as a finite combination of elements of form $x^2 y$ and uv^2. Since clearly $L_{x^2 y} R_b, L_{uv^2} R_b \in \mathrm{Rad}(\mathrm{Mul}(A))$, then

$$L_{A^3} R_J \subset \mathrm{Rad}(\mathrm{Mul}(A)).$$

This implies that

$$(L_A R_J)^3 \subset \mathrm{Rad}(\mathrm{Mul}(A)).$$

As the Jacobson radical of a Banach algebra is closed, we obtain that $\overline{L_A R_J}$ consists of quasinilpotent operators. Since $\overline{L_A R_J}$ is an ideal of $\mathrm{Mul}(A)$, then

$$\overline{L_A R_J} \subset \mathrm{Rad}(\mathrm{Mul}(A)).$$

We proved that $L_A R_J \subset \mathrm{Rad}(\mathrm{Mul}(A))$. Similarly, we have that

$$L_J R_A \subset \mathrm{Rad}(\mathrm{Mul}(A)).$$

Hence $I \subset \mathrm{Rad}(\mathrm{Mul}(A))$. Then $L_J + R_J + I$ consists of quasinilpotents if and only if $L_J/I + R_J/I$ consists of quasinilpotents in $\mathrm{Mul}(A)/I$. The last is obvious because L_a/I and R_b/I commute for every $a, b \in A$ and are quasinilpotents for every $a, b \in J$. □

Let us denote by $\mathrm{Mul}_2(A)$ the closed subalgebra of $\mathrm{Mul}(A)$ generated by all operators $L_a R_b$, where $a, b \in A$.

Corollary 5.9. *If a radical Banach algebra A is compact then*

$$\mathrm{Mul}_2(A) \subset \mathrm{Rad}(\mathrm{Mul}(A)).$$

As a consequence, the algebra $\mathcal{E\ell}(A) + \mathrm{Mul}_2(A)$ consists of quasinilpotent operators.

5.4. Permanently radical algebras

Let us call a class $\mathcal{P}$ of Banach algebras *permanent* if for each $A \in \mathcal{P}$ and each bounded homomorphism $f : A \longrightarrow B$ with dense image, the algebra B is also in $\mathcal{P}$. Examples of permanent classes are commutative algebras, separable algebras, finite-dimensional algebras, amenable algebras, algebras with bounded approximate identities.

An example of Dixon [13, Example 9.3] shows that the class of all radical Banach algebras is not permanent. We say that a Banach algebra A is *permanently radical* if for every bounded homomorphism $f : A \longrightarrow B$ the closure of the image $f(A)$ in B is a radical Banach algebra.

It follows from (i) of the following theorem that the class of all permanently radical Banach algebras is permanent. We also show that it is extension stable.

Theorem 5.10. *Let A be a Banach algebra.*

(i) *If A is permanently radical then so is $\overline{g(A)}$ for every bounded homomorphism $g : A \longrightarrow B$ of Banach algebras.*
(ii) *If a closed ideal J and the quotient A/J of A are permanently radical then A is permanently radical.*
(iii) *If $\{I_\alpha\}_{\alpha\in\Lambda}$ is an increasing net of permanently radical closed ideals in A and $\cup_{\alpha\in\Lambda} I_\alpha$ is dense in A then A is permanently radical.*

Proof. (i) Let $C = \overline{g(A)}$ and $f : C \longrightarrow D$ be a bounded homomorphism of Banach algebras with $\overline{f(C)} = D$. Then $f \circ g$ is a bounded homomorphism $A \longrightarrow D$ with dense image. If A is permanently radical then D is radical.

(ii) Let $f : A \longrightarrow B$ be a bounded homomorphism with $\overline{f(A)} = B$. Then $I := \overline{f(J)}$ is a radical ideal of B. Hence $I \subset \mathrm{Rad}(B)$, whence $g = q_{\mathrm{Rad}(B)} \circ f$ is a bounded homomorphism of A into $C = B/\mathrm{Rad}(B)$ and $I \subset \ker g$. Thus there is a bounded homomorphism $h : A/J \to C$ such that $g = h \circ q_J$. As $C = \overline{h(A/J)}$, C is radical. Then $C = 0$, whence B is radical.

(iii) If $f : A \to B$ is a bounded homomorphism with dense image then all $\overline{f(I_\alpha)}$ are radical ideals of B. Hence $f(I_\alpha) \subset \mathrm{Rad}(B)$, for each α, and $f(A) \subset \mathrm{Rad}(B)$ by density. Thus $B = \mathrm{Rad}(B)$. □

Remark 5.11. It is not clear if (ii) may be reversed. It follows from (i) that a quotient of a permanently radical Banach algebra is permanently radical, but what one can say about ideals?

Clearly the class of all permanently radical Banach algebras contains all radical commutative Banach algebras and all finite-dimensional radical algebras.

Theorem 5.12. *Every radical hypofinite Banach algebra A is permanently radical.*

Proof. Let us show first that each topologically irreducible representation π of A on a Banach space X is zero. Indeed, assume that $\pi \neq 0$ and let $J = \ker \pi$, then A/J contains a non-zero finite rank element a/J. Since $\pi(a) \neq 0$ there is $0 \neq x \in X$ with $\pi(a)x \neq 0$, whence $\pi(A)\pi(a)x$ is a dense subspace of X. Since $\pi(a)\pi(A)\pi(a)x =$

$\pi(aAa)x$ is a finite-dimensional subspace, we conclude that $\dim(\pi(a)X) < \infty$. Let $I = \pi(A) \cap \mathcal{F}(X)$, this is a non-zero ideal of $\pi(A)$. Hence I has no closed invariant subspaces. On the other hand, I consists of nilpotent operators (indeed, if a finite rank operator is the image of a quasinilpotent element under a representation then it is nilpotent). By the Lomonosov Theorem [18] (or by an earlier result of Barnes [5]), I has an invariant subspace. This contradiction shows that $\pi = 0$.

Let now $f : A \longrightarrow B$ be a continuous homomorphism with dense image. If π is a strictly irreducible representation of B then $\pi \circ f$ is a topologically irreducible representation of A whence $\pi \circ f = 0$ and $\pi = 0$. This shows that B is radical. □

Corollary 5.13. *If A is a hypofinite Banach algebra then* $\mathrm{Rad}(A)$ *is permanently radical.*

Proof. $\mathrm{Rad}(A)$ is a hypofinite Banach algebra, because it is an ideal of A (see Corollary 3.60). So apply Corollary 5.12. □

It would be convenient to formulate a result established in the proof of Theorem 5.12 as follows.

Proposition 5.14. *Each topologically irreducible representation of a radical hypofinite Banach algebra is trivial.*

Is any radical bicompact Banach algebra permanently radical? Note that the positive answer would imply that all hypocompact radical Banach algebras are permanently radical. But even if the answer is affirmative it needs another approach because the following result shows that Proposition 5.14 doesn't extend to radical hypocompact algebras.

Theorem 5.15. *There is a radical bicompact, singly generated Banach algebra A with a non-trivial topologically irreducible contractive representation by bounded operators.*

Proof. Let T be a quasinilpotent operator on a Banach space X without non-trivial closed invariant subspaces (the existence of such operators is a famous example by Read [19]). Let B be the subalgebra of $\mathcal{B}(X)$ generated by T. It follows from Bonsall's theorem [8, Theorem 3] that there is an algebra norm $\|\cdot\|'$ on B such that

1) $\|a\| \leq \|a\|'$ for each $a \in B$,
2) the completion A of B in $\|\cdot\|'$ is a Banach subalgebra of $\mathcal{B}(X)$,
3) the element b of A corresponding to T is compact.

Since A is generated by b, it is a bicompact, singly generated Banach algebra. As every compact element of a Banach algebra has countable spectrum by [1, Theorem 4.4], $\sigma_A(b) = \sigma(T)$ by Proposition 2.3(ii). Hence b is a quasinilpotent element of A, and A is radical. As A is embedded into $\mathcal{B}(X)$, let π be the natural representation of A by bounded operators on X. Then $\pi(b) = T$, and π is topologically irreducible and contractive. □

Theorem 5.16. *If A is a compact Banach algebra and* $\mathrm{Rad}(A)$ *is permanently radical, then* $L_{\mathrm{Rad}(A)} \cup R_{\mathrm{Rad}(A)} \subset \mathrm{Rad}(\mathrm{Mul}(A))$.

Proof. Let $I = \mathrm{Rad}(\mathrm{Mul}(A))$, $C = \mathrm{Mul}(A)/I$ and $q = q_I$. Define $\phi : \mathrm{Rad}(A) \to C$ by $\phi(a) = q(L_a)$ for any a. Then the algebra $\overline{\phi(\mathrm{Rad}(A))}$ is radical.

For any $a \in \mathrm{Rad}(A)$ and $T \in \mathcal{E\ell}(A)$, we have that

$$L_a T \in L_{\mathrm{Rad}(A)} + L_{\mathrm{Rad}(A)} R_A \subset L_{\mathrm{Rad}(A)} + \mathrm{Rad}(\mathrm{Mul}(A))$$

by Theorem 5.8. It follows that $q(L_a T) \in \overline{\phi(\mathrm{Rad}(A))}$. By continuity, the same is true for all $T \in \mathrm{Mul}(A)$. Thus all $q(L_a T)$ are quasinilpotent. This shows that $L_{\mathrm{Rad}(A)}\mathrm{Mul}(A)$ consists of quasinilpotents, whence $L_{\mathrm{Rad}(A)} \subset \mathrm{Rad}(\mathrm{Mul}(A))$. Similarly, we have that $R_{\mathrm{Rad}(A)} \subset \mathrm{Rad}(\mathrm{Mul}(A))$. □

Corollary 5.17. *If A is an approximable Banach algebra then*

$$L_{\mathrm{Rad}(A)} \cup R_{\mathrm{Rad}(A)} \subset \mathrm{Rad}(\mathrm{Mul}(A)).$$

Proof. Clearly A is compact. Furthermore, $\mathrm{Rad}(A)$ is permanently radical by Corollary 5.13. □

Corollary 5.18. *If A is an approximable Banach algebra and A is commutative modulo* $\mathrm{Rad}(A)$*, then* $\mathrm{Mul}(A)$ *is commutative modulo* $\mathrm{Rad}(\mathrm{Mul}(A))$.

Proof. For all $a, b \in A$, $[L_a, L_b] = L_{[a,b]} \in L_{\mathrm{Rad}(A)} \subset \mathrm{Rad}(\mathrm{Mul}(A))$ and, similarly, $[R_a, R_b] \in \mathrm{Rad}(\mathrm{Mul}(A))$. Since also $[L_a, R_b] = 0 \in \mathrm{Rad}(\mathrm{Mul}(A))$, we get that $\mathrm{Mul}(A)/\mathrm{Rad}(\mathrm{Mul}(A))$ is commutative. □

5.5. Chains of closed ideals

Now we consider invariant subspaces of the algebras of elementary operators. It was proved by Wojtyński [36] that the well-known problem of the existence of a non-trivial closed ideal in a radical Banach algebra has the positive answer if the algebra has a non-zero compact element. The proof of this fact, based on the invariant subspace theorem for Volterra semigroups is given in [31]. The following theorem presents another proof and a slightly more general formulation of this result.

Recall that a *central multiplier* on a Banach algebra A is a bounded linear operator on A commuting with left and right multiplications.

Theorem 5.19. *If a radical Banach algebra A has a non-zero compact element then either the multiplication in A is trivial or A has a closed ideal invariant under all central multipliers.*

Proof. Let an element $a \in A$ be compact. Then $I = \{b : L_a R_b \in \mathcal{K}(A)\}$ is a non-zero closed ideal in A invariant under central multipliers. So we have to assume that $I = A$. Setting $J = \{c : L_c R_b \in \mathcal{K}(A)\}$ for all $b \in A$, we similarly reduce to the case that $J = A$. In other words, we may suppose that A is bicompact.

Recall that a subspace invariant under an algebra of operators and its commutant is called *hyperinvariant* for this algebra. Note that the set of all central multipliers is the commutant of Mul(A). So our aim is to show that Mul(A) has a non-trivial hyperinvariant subspace. Since $\mathrm{Mul}_2(A)$ is an ideal in Mul(A) it suffices to show the same for $\mathrm{Mul}_2(A)$ (see for example [31]). By Corollary 5.9, $\mathrm{Mul}_2(A)$ is a radical algebra of compact operators. Hence it has a hyperinvariant subspace by [24]. □

Is this possible to strengthen the result and to prove the existence of a total chain of closed ideals? We will show that the answer to this question is negative.

Recall that a chain (i.e., a set linearly ordered by the inclusion) $\mathfrak{N}$ of closed subspaces of a Banach space X is *total* if it is not contained in a larger chain of subspaces. This is equivalent to the conditions that $\mathfrak{N}$ is complete and for any elements $Y_1 \subset Y_2$ of $\mathfrak{N}$, either $\dim(L_2/L_1) = 1$ or there exists an intermediate subspace in $\mathfrak{N}$.

Let us call by *a gap* in the lattice of closed ideals of a Banach algebra A a pair $I_1 \subset I_2$ of closed ideals without intermediate ideals, and in this case the quotient I_2/I_1 is called a *gap-quotient* of the lattice. It is easy to show by transfinite induction that if $\dim(I_2/I_1) = 1$ for any gap, then A has a total chain of closed ideals (moreover, each chain of ideals extends to a total one).

An example of a gap is a pair $(0, I)$ where I is a minimal closed ideal. So if each chain of closed ideals in A extends to a total one then each minimal closed ideal is one-dimensional. We show now that these properties can fail in the class of radical bicompact algebras. Then it will be shown that for radical hypofinite algebras the situation is different.

Theorem 5.20. (i) *There is a radical bicompact Banach algebra without a total chain of closed ideals.*

(ii) *A radical bicompact Banach algebra can have an infinite-dimensional minimal closed ideal.*

Proof. Let A be a commutative bicompact radical Banach algebra with a topologically irreducible representation $\pi : A \longrightarrow \mathcal{B}(X)$ (see Theorem 5.15). On the Banach space $B = A \oplus X$ with the norm $\|a \oplus x\| = \max\{\|a\|, \|x\|\}$ introduce a multiplication by $(a \oplus x)(b \oplus y) = ab \oplus \pi(a)y$. Then B is a Banach algebra. Since $(a \oplus x)^n = a^n \oplus \pi(a^{n-1})x$, then $\|(a \oplus x)^n\| \leq (\|a\| + \|x\|)\|a^{n-1}\|$ for every $n > 0$, whence B is radical.

We show that B is a bicompact algebra. For any $a \oplus x, b \oplus y \in B$, the operator $T = L_{a \oplus x} R_{b \oplus y}$ maps any $c \oplus z$ into $acb \oplus \pi(ac)y$. As $\mathrm{ball}(B) = \mathrm{ball}(A) \oplus \mathrm{ball}(X)$, we obtain that

$$T(\mathrm{ball}(B)) \subset L_a R_b(\mathrm{ball}(A)) \oplus \pi(L_a(\mathrm{ball}(A))y).$$

As all operators $L_a R_b$ in A are compact, it suffices to prove the precompactness of the set $\pi(L_a(\mathrm{ball}(A))y)$. In other words, we have to show that any operator $S_y : c \longmapsto \pi(ac)y$ is compact. If take $y \in \pi(A)X$ with $y = \pi(d)z$ for some $d \in A$ and $z \in X$, then S_y is compact because it decomposes through $L_a R_d$. It follows that S_y is compact for any y in the linear span Y of $\pi(A)X$. But Y is dense in X

because it is invariant for $\pi(A)$. Hence for any $y \in X$ there is a sequence $y_n \to y$ in Y. It follows that $\|S_y - S_{y_n}\| \to 0$, so S_y is compact for every $y \in X$.

The subspace $I = 0 \oplus X$ is a closed ideal of B and it follows easily from topological transitivity of π that I is a minimal closed ideal. Moreover, each non-zero closed ideal J of B contains I. Indeed, J cannot be a subspace of I. Hence there is $a \oplus x \in J$ with $a \neq 0$. But then

$$0 \oplus \pi(b)\pi(a)y = (b \oplus 0)(a \oplus x)(0 \oplus y) \in J$$

for any $b \in A$, $y \in X$, whence $I \subset J$.

We see that B has no total chains of closed ideals and has an infinite-dimensional minimal closed ideal. □

In the remaining part of the section we obtain some "affirmative" results.

Lemma 5.21. *If A is a radical compact Banach algebra and $J \subset I$ is a gap of closed ideals of A then $AI \subset J$ or $IA \subset J$.*

Proof. Assume the contrary. Then $\overline{AI} = \overline{IA} = I$ (otherwise we obtain an intermediate ideal) whence $\overline{AIA} = I$. One may assume that $\dim(I/J) > 1$ because otherwise the statement is trivial. Let π be the natural representation of $\mathrm{Mul}(A)$ on the space $X = I/J$. It is topologically irreducible because if Y is an invariant closed subspace of π then $\{x \in I : x/J \in Y\}$ is a closed ideal between J and I. By Corollary 5.9, the algebra $\mathrm{Mul}_2(A)$ is contained in the radical of $\mathrm{Mul}(A)$. If $\pi(\mathrm{Mul}_2(A))$ contains a non-zero compact operator then it has a non-trivial invariant closed subspace by the Lomonosov Theorem [18]. As $\pi(\mathrm{Mul}_2(A))$ is an ideal of $\pi(\mathrm{Mul}(A))$, this implies that $\pi(\mathrm{Mul}(A))$ has a non-trivial invariant closed subspace, a contradiction.

As we saw in the proof of Theorem 5.8, the operator $L_{a^2}R_b$ is a compact operator in $\mathrm{Mul}_2(A)$ for every $a, b \in A$. Therefore, $\pi(L_{a^2}R_b)$ is also a compact operator. By the above,

$$\pi(L_{a^2}R_b) = 0.$$

In other words, $a^2Ib \subset J$. Since $\overline{IA} = I$, we get that $a^2I \subset J$. Thus

$$\pi(L_a)^2 = 0$$

for all $a \in A$, whence $A^3I \subset J$ by the Nagata–Higman theorem. Since $\overline{AI} = I$, we obtain a contradiction. □

Theorem 5.22. *If A is an infinite-dimensional compact radical Banach algebra then any chain of closed ideals of A extends to an infinite chain of closed ideals of A.*

Proof. Suppose, to the contrary, that there is a maximal chain

$$0 = J_0 \subset J_1 \subset \cdots \subset J_n = A$$

of closed ideals of A. Then each pair (J_{k-1}, J_k) is a gap. It follows from Lemma 5.21 that $AJ_kA \subset J_{k-1}$ for every $k > 0$. Hence

$$A^{2n+1} = 0.$$

It follows that $\mathrm{Mul}(A)$ is also a nilpotent algebra. Hence it has no non-trivial topologically irreducible representations. But for any gap (J_{k-1}, J_k) its representation on $X_k = J_k/J_{k-1}$ is topologically irreducible and at least one of X_k must be infinite dimensional. We obtained a contradiction. □

Recall that by $\mathcal{A}(X)$ we denote the operator norm closure of the ideal $\mathcal{F}(X)$ of all finite rank operators on X. Our aim is to show that if $\mathcal{A}(X) \neq \mathcal{K}(X)$ then between $\mathcal{A}(X)$ and $\mathcal{K}(X)$ there are intermediate closed ideals.

Corollary 5.23. (i) *If* $\dim(\mathcal{K}(X)/\mathcal{A}(X)) = n$ *(where n is a finite number or ∞) then $\mathcal{K}(X)$ has a chain of n different closed ideals, containing $\mathcal{F}(X)$.*

(ii) *Let M and N be closed ideals of $\mathcal{B}(X)$ with $\mathcal{A}(X) \subset N \subsetneqq M \subset \mathcal{K}(X)$. If M^2 is not contained in N then $\mathcal{B}(X)$ has a closed ideal between N and M. In particular, if $(\mathcal{K}(X)/\mathcal{A}(X))^2 \neq 0$ then there is a closed ideal of $\mathcal{B}(X)$ between $\mathcal{A}(X)$ and $\mathcal{K}(X)$.*

(iii) *If the algebra $\mathcal{K}(X)/\mathcal{A}(X)$ is not nilpotent then every maximal chain of closed ideals of $\mathcal{B}(X)$ between $\mathcal{A}(X)$ and $\mathcal{K}(X)$ is infinite.*

Proof. (i) The algebra $\mathcal{Q}(X) = \mathcal{K}(X)/\mathcal{A}(X)$ is radical by Corollary 3.54. If its dimension n is finite then clearly it has a chain of n ideals (since the nilpotent algebra $\mathrm{Mul}(\mathcal{Q}(X))$ is triangularizable). In any case it is bicompact, so if $n = \infty$ then it has an infinite chain of ideals by Theorem 5.22. The preimages of these ideals in $\mathcal{K}(X)$ form a chain of ideals of $\mathcal{K}(X)$ containing $\overline{\mathcal{F}(X)}$. This proves (i).

(ii) Assume, to the contrary, that there are no closed ideals between N and M. As $\overline{M^2 + N}$ is a closed ideal of $\mathcal{B}(X)$ strictly containing N, then

$$M = \overline{M^2 + N}.$$

As $\{T \in M : TM \subset N\}$ is a closed ideal of $\mathcal{B}(X)$ strictly contained in M, then

$$N = \{T \in M : TM \subset N\}$$

and, similarly,

$$N = \{T \in M : MT \subset N\}.$$

By (i), there is a closed ideal I of $\mathcal{K}(X)$ intermediate between M and N. Set $J = \overline{MIM + N}$. Then

$$N \subset J \subset I \subsetneqq M.$$

If $J = N$ then $MIM \subset N$ whence, by above, $IM \subset N$ and therefore $I \subset N$, a contradiction. Thus

$$N \subsetneqq J \subsetneqq M.$$

As

$$\mathcal{B}(X)J\mathcal{B}(X) \subset \overline{\mathcal{B}(X)MIM\mathcal{B}(X) + N} \subset \overline{MIM + N} = J,$$

we obtained that J is an intermediate closed ideal of $\mathcal{B}(X)$ between N and M. Part (ii) is proved.

(iii) Assuming that $\mathcal{Q}(X)$ is not nilpotent, choose a maximal chain (I_α) of closed ideals of $\mathcal{B}(X)$ between $\mathcal{A}(X)$ and $\mathcal{K}(X)$. If it is finite, namely

$$\mathcal{A}(X) = I_0 \subsetneqq I_1 \subsetneqq I_2 \subsetneqq \cdots \subsetneqq I_n = \mathcal{K}(X),$$

then

$$I_k^2 \subset I_{k-1}$$

for every $k > 0$ by (ii). Then $\mathcal{Q}(X)^{2^n} = 0$, a contradiction. □

Example 5.24. *To construct an example of a Banach space X for which the algebra $\mathcal{K}(X)/\mathcal{A}(X)$ is not nilpotent, one can use a remarkable result of Willis [34].*

Recall that X is said to have approximation property (*AP*) (*respectively,* compact approximation property (*CAP*)) *if for each compact set $M \subset X$ and each $\varepsilon > 0$, there is a finite rank (respectively, compact) operator $S = S(M,\varepsilon)$ with $\|Sx - x\| < \varepsilon$ for all $x \in M$. If S always can be chosen in such a way that $\|S\| \le C$ for some fixed $C > 0$ then one says that X has* bounded approximation property (*BAP*) (*respectively,* bounded compact approximation property (*BCAP*)).

It was proved in [34] that there exists a space X which has not AP but has BCAP. Let us show that this is a space we need. Indeed, it follows from BCAP that the algebra $\mathcal{K}(X)$ has a bounded approximate identity: to construct it one have to take for the index set the set of all pairs $\lambda = (M,\varepsilon)$ where M is a compact subset of X and $\varepsilon > 0$, and denote by S_λ an operator $S = S(M,\varepsilon)$ from the definition of BCAP. In particular, $\mathcal{K}(X)^n$ is dense in $\mathcal{K}(X)$ for each n. Hence if $\mathcal{K}(X)/\mathcal{A}(X)$ is nilpotent then $\mathcal{A}(X) = \mathcal{K}(X)$. Therefore $\mathcal{A}(X)$ has a bounded approximate identity e_λ and one can assume that $e_\lambda \in \mathcal{F}(X)$ for each λ. Let us show that this implies AP (in contradiction with the choice of X).

It is easy to see (considering rank one operators) that $e_\lambda x \to x$ for each $x \in X$. Now if a compact subset M of X and $\varepsilon > 0$ are given, let us choose a finite ε-net M_0 in M and an index μ with $\|e_\mu x - x\| < \varepsilon$ for all $x \in M_0$. Then $\|e_\mu x - x\| < t\varepsilon$ for all $x \in M$, where $t = 2 + \sup_\lambda \|e_\lambda\|$.

Let us denote by $\mathcal{A}$ and $\mathcal{K}$ the closed operator ideals of approximable and compact operators, respectively.

Corollary 5.25. *There is an infinite chain of closed operator ideals intermediate between $\mathcal{A}$ and $\mathcal{K}$.*

Proof. Let Z be a Banach space with non-nilpotent $\mathcal{K}(Z)/\mathcal{A}(Z)$ (see Example 5.24). By Corollary 5.23, between $\mathcal{K}(Z)$ and $\mathcal{A}(Z)$ there is an infinite chain $\{I_\alpha\}$ of closed ideals of $\mathcal{B}(Z)$. For each pair (X,Y) of Banach spaces, we denote by $\mathcal{U}_\alpha(X,Y)$ the set of all operators $T \in \mathcal{K}(X,Y)$ such that $ATB \in I_\alpha$ for all $A \in \mathcal{B}(Y,Z)$ and $B \in \mathcal{B}(Z,X)$. It is easy to check that each $\mathcal{U}_\alpha$ is a closed operator ideal between $\mathcal{A}$ and $\mathcal{K}$, that all $\mathcal{U}_\alpha$ are different and that they form a chain. □

Theorem 5.26. *If A is a radical approximable Banach algebra then each gap-quotient in the lattice of the closed ideals of A is one-dimensional.*

Proof. Let $J \subset I$ be a gap of closed ideals of A. Then either $AI \subset J$ or $IA \subset J$ by Lemma 5.21. Suppose that $IA \subset J$. Denote by π the natural representation of $\mathrm{Mul}(A)$ on I/J. Then we have that $\pi(R_A) = 0$, whence $\pi(\mathcal{E}\ell(A)) = \pi(L_A)$ and $\pi(\mathrm{Mul}(A)) \subset \overline{\pi(L_A)}$.

The map $a \longmapsto \pi(L_a)$ is a topologically irreducible representation of A on I/J. Since such a representation of A must be trivial by Proposition 5.14, it acts on a one-dimensional space. □

Corollary 5.27. *Every radical hypofinite Banach algebra has a total chain of closed ideals, and each minimal closed ideal in such an algebra is one-dimensional.*

Let us call a subspace I of a Banach algebra A *a quasiideal* if $AIA \subset I$. Clearly each ideal is a quasiideal. The converse is true if A has a (non-necessarily bounded) approximate identity.

Theorem 5.28. *Any bicompact radical Banach algebra has a total chain of closed quasiideals.*

Proof. Closed quasiideals are invariant subspaces of the radical algebra $\mathrm{Mul}_2(A)$ of compact operators. Since such algebras are triangularizable, our statement follows. □

6. Spectral subspaces of elementary and multiplication operators

In this section we consider invariant subspaces of semicompact multiplication operators, on which the operators are surjective (in particular, eigenspaces with non-zero eigenvalues or spectral subspaces corresponding to clopen subsets of spectra non-containing 0). Our approach will be based (apart of the tensor radical technique) on a study of operators acting in ordered pairs of Banach spaces. In Section 6.4 we improve the results for semicompact elementary operators by another technique to show that such invariant subspaces are contained in the component of every quasi-Banach operator ideal.

6.1. Operators on an ordered pair of Banach spaces

Let $\mathcal{X}, \mathcal{Y}$ be Banach spaces, and $\mathcal{Y} \subset \mathcal{X}$. Suppose that

$$\|y\|_{\mathcal{X}} \leq \|y\|_{\mathcal{Y}} \tag{6.1}$$

for all $y \in \mathcal{Y}$. We refer to such a subspace $\mathcal{Y}$ as a *Banach subspace* of $\mathcal{X}$ and call $(\mathcal{Y}, \mathcal{X})$ an *ordered pair of Banach spaces.*

Denote by $\mathcal{B}(\mathcal{X}||\mathcal{Y})$ the space of all operators $T \in \mathcal{B}(\mathcal{X})$ such that $T\mathcal{Y} \subset \mathcal{Y}$. It is non-zero, for instance the identity operator $1_{\mathcal{X}}$ in $\mathcal{B}(\mathcal{X})$ lies in $\mathcal{B}(\mathcal{X}||\mathcal{Y})$.

Theorem 6.1. *Let $\mathcal{Y}$ be a Banach subspace of a Banach space $\mathcal{X}$. Then $T|_{\mathcal{Y}} \in \mathcal{B}(\mathcal{Y})$ for any $T \in \mathcal{B}(\mathcal{X}||\mathcal{Y})$, and $\mathcal{B}(\mathcal{X}||\mathcal{Y})$ is a unital Banach subalgebra of $\mathcal{B}(\mathcal{X})$ with respect to the norm*

$$\|T\|_{\mathcal{B}(\mathcal{X}||\mathcal{Y})} = \max\left\{\|T\|_{\mathcal{B}(\mathcal{X})}, \|T|_{\mathcal{Y}}\|_{\mathcal{B}(\mathcal{Y})}\right\}$$

for any $T \in \mathcal{B}(\mathcal{X}||\mathcal{Y})$.

Proof. We first show that $T|_\mathcal{Y}$ is a bounded operator on $\mathcal{Y}$. To apply the Closed Graph Theorem, it is sufficient to show that the conditions $\|y_n\|_\mathcal{Y} \to 0$ and $\|Ty_n - u\|_\mathcal{Y} \to 0$ as $n \to \infty$ for $\{y_n\} \subset \mathcal{Y}$ imply $u = 0$. If these conditions hold, then also $\|y\|_\mathcal{X} \to 0$ and $\|Ty_n - u\|_\mathcal{X} \to 0$ as $n \to \infty$. As T is bounded on $\mathcal{X}$, then $u = 0$.

As a consequence, $\|\cdot\|_{\mathcal{B}(\mathcal{X}||\mathcal{Y})}$ is a norm on $\mathcal{B}(\mathcal{X}||\mathcal{Y})$. This norm is clearly a unital algebra norm that majorizes $\|\cdot\|_{\mathcal{B}(\mathcal{X})}$ on $\mathcal{B}(\mathcal{X}||\mathcal{Y})$. To finish the proof, it remains to show that $\|\cdot\|_{\mathcal{B}(\mathcal{X}||\mathcal{Y})}$ is complete.

Let $\{T_n\}$ be a fundamental sequence in $\mathcal{B}(\mathcal{X}||\mathcal{Y})$. Then there are $T \in \mathcal{B}(\mathcal{X})$ and $S \in \mathcal{B}(\mathcal{Y})$ such that $\|T_n - T\|_{\mathcal{B}(\mathcal{X})} \to 0$ and $\|T_n|_\mathcal{Y} - S\|_{\mathcal{B}(\mathcal{Y})} \to 0$ as $n \to \infty$. Then $\|T_n y - Ty\|_\mathcal{X} \to 0$ and $\|T_n y - Sy\|_\mathcal{X} \le \|T_n y - Sy\|_\mathcal{Y} \to 0$ as $n \to \infty$, for every $y \in \mathcal{Y}$. This shows that $\mathcal{Y}$ is invariant for T and $T|_\mathcal{Y} = S$. □

As usual, $\|S\|_{\mathcal{B}(\mathcal{X},\mathcal{Y})}$ denotes the operator norm of an operator S in $\mathcal{B}(\mathcal{X},\mathcal{Y})$.

Proposition 6.2. *Let $\mathcal{Y}$ be a Banach subspace of a Banach space $\mathcal{X}$. Then $\mathcal{B}(\mathcal{X},\mathcal{Y})$ is a Banach algebra with respect to the usual norm $\|\cdot\|_{\mathcal{B}(\mathcal{X},\mathcal{Y})}$.*

Proof. It is clear that $\mathcal{B}(\mathcal{X},\mathcal{Y})$ is a Banach space. If $S, T \in \mathcal{B}(\mathcal{X},\mathcal{Y})$ then

$$\|STx\|_\mathcal{Y} \le \|S\|_{\mathcal{B}(\mathcal{X},\mathcal{Y})}\|Tx\|_\mathcal{X} \le \|S\|_{\mathcal{B}(\mathcal{X},\mathcal{Y})}\|Tx\|_\mathcal{Y} \le \|S\|_{\mathcal{B}(\mathcal{X},\mathcal{Y})}\|T\|_{\mathcal{B}(\mathcal{X},\mathcal{Y})}\|x\|_\mathcal{X}$$

for every $x \in \mathcal{X}$, whence $\mathcal{B}(\mathcal{X},\mathcal{Y})$ is a Banach algebra. □

Proposition 6.3. *Let $\mathcal{Y}$ be a Banach subspace of a Banach space $\mathcal{X}$. Then $\mathcal{B}(\mathcal{X},\mathcal{Y})$ is a Banach ideal of $\mathcal{B}(\mathcal{X}||\mathcal{Y})$ with respect to $\|\cdot\|_{\mathcal{B}(\mathcal{X},\mathcal{Y})}$ which is a flexible norm.*

Proof. The inclusion $\mathcal{B}(\mathcal{X},\mathcal{Y}) \subset \mathcal{B}(\mathcal{X}||\mathcal{Y})$ follows by Theorem 6.1. Let $S \in \mathcal{B}(\mathcal{X},\mathcal{Y})$ and $P, T \in \mathcal{B}(\mathcal{X}||\mathcal{Y})$. It is clear that $PS, ST \in \mathcal{B}(\mathcal{X},\mathcal{Y}) \subset \mathcal{B}(\mathcal{X}||\mathcal{Y})$. So $\mathcal{B}(\mathcal{X},\mathcal{Y})$ is an ideal of $\mathcal{B}(\mathcal{X}||\mathcal{Y})$. As $\|Sx\|_\mathcal{Y} \le \|S\|_{\mathcal{B}(\mathcal{X},\mathcal{Y})}\|x\|_\mathcal{X}$ for every $x \in \mathcal{X}$, one obtains from (6.1) that

$$\begin{aligned} &\|Sx\|_\mathcal{X} \le \|Sx\|_\mathcal{Y} \le \|S\|_{\mathcal{B}(\mathcal{X},\mathcal{Y})}\|x\|_\mathcal{X} \text{ and} \\ &\|Sy\|_\mathcal{Y} \le \|S\|_{\mathcal{B}(\mathcal{X},\mathcal{Y})}\|y\|_\mathcal{X} \le \|S\|_{\mathcal{B}(\mathcal{X},\mathcal{Y})}\|y\|_\mathcal{Y} \end{aligned}$$

for every $x \in \mathcal{X}$ and $y \in \mathcal{Y}$. Therefore

$$\|S\|_{\mathcal{B}(\mathcal{X})} \le \max\left\{\|S\|_{\mathcal{B}(\mathcal{X})}, \|S|_\mathcal{Y}\|_{\mathcal{B}(\mathcal{Y})}\right\} = \|S\|_{\mathcal{B}(\mathcal{X}||\mathcal{Y})} \le \|S\|_{\mathcal{B}(\mathcal{X},\mathcal{Y})}.$$

It follows that

$$\begin{aligned} \|PSTx\|_\mathcal{Y} &\le \|P|_\mathcal{Y}\|_{\mathcal{B}(\mathcal{Y})}\,\|STx\|_\mathcal{Y} \le \|P|_\mathcal{Y}\|_{\mathcal{B}(\mathcal{Y})}\,\|S\|_{\mathcal{B}(\mathcal{X},\mathcal{Y})}\,\|Tx\|_\mathcal{X} \\ &\le \|P|_\mathcal{Y}\|_{\mathcal{B}(\mathcal{Y})}\,\|S\|_{\mathcal{B}(\mathcal{X},\mathcal{Y})}\,\|T\|_{\mathcal{B}(\mathcal{X})}\,\|x\|_\mathcal{X} \\ &\le \|P\|_{\mathcal{B}(\mathcal{X}||\mathcal{Y})}\,\|S\|_{\mathcal{B}(\mathcal{X},\mathcal{Y})}\,\|T\|_{\mathcal{B}(\mathcal{X}||\mathcal{Y})}\,\|x\|_\mathcal{X}\,. \end{aligned}$$

for every $x \in \mathcal{X}$, whence

$$\|PST\|_{\mathcal{B}(\mathcal{X},\mathcal{Y})} \le \|P\|_{\mathcal{B}(\mathcal{X}||\mathcal{Y})}\,\|S\|_{\mathcal{B}(\mathcal{X},\mathcal{Y})}\,\|T\|_{\mathcal{B}(\mathcal{X}||\mathcal{Y})}\,.$$

□

Theorem 6.4. *Let $\mathcal{Y}$ be a Banach subspace of a Banach space $\mathcal{X}$. Then every operator $T \in \mathcal{B}(\mathcal{X})$ such that $T\mathcal{X} \subset \mathcal{Y}$ belongs to $\mathcal{B}(\mathcal{X}, \mathcal{Y})$.*

Proof. By the Closed Graph Theorem, it suffices to show that $u = 0$ if $\|x_n\|_{\mathcal{X}} \to 0$ and $\|Tx_n - u\|_{\mathcal{Y}} \to 0$ as $n \to \infty$. Indeed, the last implies that $\|Tx_n - u\|_{\mathcal{X}} \to 0$ as $n \to \infty$. As T is bounded on $\mathcal{X}$, then $u = 0$. □

6.2. Invariant subspaces for operators on an ordered pair of Banach spaces

We consider those invariant subspaces of an operator on an ordered pair $(\mathcal{Y}, \mathcal{X})$ of Banach spaces on which the operator is surjective. Clearly such a subspace is contained in $\mathcal{Y}$ if the operator belongs to $\mathcal{B}(\mathcal{X}, \mathcal{Y})$. We show that the same is true if the operator belongs to $\mathcal{B}(\mathcal{X}\|\mathcal{Y})$ and is quasinilpotent modulo $\mathcal{B}(\mathcal{X}, \mathcal{Y})$.

Theorem 6.5. *Let $\mathcal{X}$ be a Banach space, $\mathcal{Y}$ a Banach subspace of $\mathcal{X}$, and let $T \in \mathcal{B}(\mathcal{X}\|\mathcal{Y})$. Assume that*

$$T \in Q_{\mathcal{B}(\mathcal{X},\mathcal{Y})}(\mathcal{B}(\mathcal{X}\|\mathcal{Y})). \tag{6.2}$$

If $\mathcal{Z}$ is a closed subspace of $\mathcal{X}$ such that $\mathcal{Z} = T\mathcal{Z}$, then $\mathcal{Z} \subset \mathcal{Y}$. Moreover, the norms $\|\cdot\|_{\mathcal{X}}$ and $\|\cdot\|_{\mathcal{Y}}$ are equivalent on $\mathcal{Z}$, so $\mathcal{Z}$ is also closed in $\mathcal{Y}$.

Proof. By the Open Mapping Theorem, there is $t > 0$ such that for each $z \in \mathcal{Z}$ there is $w \in \mathcal{Z}$ with $Tw = z$ and

$$\|w\|_{\mathcal{X}} \le t\|z\|_{\mathcal{X}}.$$

Let $\varepsilon > 0$ be such that $\varepsilon < t^{-1}$. It follows from Proposition 6.3 that $\mathcal{B}(\mathcal{X}, \mathcal{Y})$ is an ideal of $\mathcal{B}(\mathcal{X}\|\mathcal{Y})$. By our assumption and Proposition 2.1(iii), there is $m \in \mathbb{N}$ such that

$$\operatorname{dist}_{\mathcal{B}(\mathcal{X}\|\mathcal{Y})}(T^m, \mathcal{B}(\mathcal{X}, \mathcal{Y})) < \varepsilon^m.$$

Therefore there is an operator $S \in \mathcal{B}(\mathcal{X}, \mathcal{Y})$ and an operator $P \in \mathcal{B}(\mathcal{X}\|\mathcal{Y})$ such that

$$T^m = S + P \text{ with } \max\left\{\|P\|_{\mathcal{B}(\mathcal{X})}, \|P|_{\mathcal{Y}}\|_{\mathcal{B}(\mathcal{Y})}\right\} < \varepsilon^m.$$

Then

$$\|Py\|_{\mathcal{Y}} \le \varepsilon^m \|y\|_{\mathcal{Y}} \tag{6.3}$$

for every $y \in \mathcal{Y}$.

It follows from the definition of t that for every $z \in \mathcal{Z}$, there is $z^\circ \in \mathcal{Z}$ with $T^m z^\circ = z$ and

$$\|z^\circ\|_{\mathcal{X}} < t^m\|z\|_{\mathcal{X}}.$$

Let $z_0 := z \in \mathcal{Z}$ be arbitrary. Set $z_1 = z_0^\circ$, $z_2 = z_1^\circ$, and so on:

$$z_{k+1} = z_k^\circ$$

for every integer $k > 0$. Thus $z_k = T^m z_{k+1} = Sz_{k+1} + Pz_{k+1}$ with

$$\|z_k\|_{\mathcal{X}} \le t^{mk}\|z\|_{\mathcal{X}}. \tag{6.4}$$

Rewriting this in the form

$$z_k - Pz_{k+1} = Sz_{k+1},$$

multiplying both sides of the equation by P^k and summing obtained equalities for $k = 0, 1, 2, \ldots$, one formally obtains that

$$z = Sz_1 + PSz_2 + P^2Sz_3 + \cdots . \tag{6.5}$$

Since $S\mathcal{X} \subset \mathcal{Y}$, all elements Sz_k and $P^{k-1}Sz_k$ belong to $\mathcal{Y}$. As $\|z_k\|_{\mathcal{X}} < t^{mk}\|z\|_{\mathcal{X}}$, one obtains that

$$\|Sz_k\|_{\mathcal{Y}} < \|S\|_{\mathcal{B}(\mathcal{X},\mathcal{Y})} \|z_k\|_{\mathcal{X}} \le t^{mk}\|S\|_{\mathcal{B}(\mathcal{X},\mathcal{Y})} \|z\|_{\mathcal{X}} . \tag{6.6}$$

It follows from (6.3) and (6.6) that

$$\left\|P^{k-1}Sz_k\right\|_{\mathcal{Y}} \le \varepsilon^{m(k-1)} \|Sz_k\|_{\mathcal{Y}} \le (\varepsilon t)^{mk} \left(\varepsilon^{-m}\|S\|_{\mathcal{B}(\mathcal{X},\mathcal{Y})} \|z\|_{\mathcal{X}}\right).$$

As $\varepsilon t < 1$, we have that

$$\sum_{k=1}^{\infty} \left\|P^{k-1}Sz_k\right\|_{\mathcal{Y}} \le \left(\frac{t^m}{1-\varepsilon^m t^m}\|S\|_{\mathcal{B}(\mathcal{X},\mathcal{Y})}\right) \|z\|_{\mathcal{X}} < \infty. \tag{6.7}$$

Since $\|\cdot\|_{\mathcal{Y}}$ is a complete norm on $\mathcal{Y}$, it follows from (6.7) and (6.5) that $z \in \mathcal{Y}$ with the estimation

$$\|z\|_{\mathcal{Y}} \le \left(\frac{t^m}{1-\varepsilon^m t^m}\|S\|_{\mathcal{B}(\mathcal{X},\mathcal{Y})}\right) \|z\|_{\mathcal{X}} .$$

Therefore $\|\cdot\|_{\mathcal{X}}$ and $\|\cdot\|_{\mathcal{Y}}$ are equivalent on $\mathcal{Z}$. As $\mathcal{Z}$ is closed with respect to $\|\cdot\|_{\mathcal{X}}$, it is closed with respect to $\|\cdot\|_{\mathcal{Y}}$. □

Corollary 6.6. *Let $\mathcal{X}, \mathcal{Y}$ and T be as in Theorem 6.5. Then*

(i) *If $\lambda \neq 0$ is an eigenvalue of T then the eigenspace $\{x \in \mathcal{X} : Tx = \lambda x\}$ is contained in $\mathcal{Y}$.*

(ii) *If σ_0 is a clopen subset of the spectrum $\sigma_{\mathcal{B}(\mathcal{X})}(T)$ of T and $0 \notin \sigma_0$ then the spectral subspace $E_{\sigma_0}(T)$ is contained in $\mathcal{Y}$.*

Proof. Indeed, these subspaces are closed in $\mathcal{X}$, invariant under T, and the restriction of T to everyone of them is invertible. □

6.3. Semicompact multiplication operators

In this section we apply Theorem 6.5 to semicompact multiplication operators considering their action on an ordered pair of spaces of nuclear and, respectively, bounded operators.

6.3.1. Multiplication operators on an ordered pair of operator ideals. First we estimate the norms of multiplication operators on an ordered pair of components of Banach operator ideals.

Let $V = \mathcal{V}(X,Y)$ and $U = \mathcal{U}(X,Y)$, where $\mathcal{V}$ and $\mathcal{U}$ are Banach operator ideals. We assume that $V \subset U$ and that V is a Banach subspace of U. As V is an invariant subspace for the algebra $\widehat{\mathcal{B}}_*(U)$ of all multiplication operators on U, then the algebra $\mathcal{B}(\mathcal{X}\|\mathcal{Y})$ contains $\widehat{\mathcal{B}}_*(U)$ by Theorem 6.1.

Lemma 6.7. *Let $V = \mathcal{V}(X,Y)$ and $U = \mathcal{U}(X,Y)$ for Banach operator ideals $\mathcal{V}$ and $\mathcal{U}$, and $V \subset U$. Then $\|T|_V\|_{\widehat{\mathcal{B}}_*(V)} \le \|T\|_{\widehat{\mathcal{B}}_*(U)}$ for every $T \in \widehat{\mathcal{B}}_*(U)$.*

Proof. Let $W = \mathcal{B}(Y) \widehat{\otimes} \mathcal{B}(X)^{\mathrm{op}}$ for brevity. Recall that the norms $\|\cdot\|_{\widehat{\mathcal{B}}_*(V)}$ in $\widehat{\mathcal{B}}_*(V)$ and $\|\cdot\|_{\widehat{\mathcal{B}}_*(U)}$ in $\widehat{\mathcal{B}}_*(U)$ are the quotient norms inherited from $W/\ker\psi$ and $W/\ker\varphi$ respectively, where $\psi : W \longrightarrow \mathcal{B}(V)$ and $\varphi : W \longrightarrow \mathcal{B}(U)$ are bounded homomorphisms that associate with every $a \otimes b \in W$ the operator $L_a R_b$.

Let $P = \varphi(w)$ and $S = \psi(w)$ for some $w \in W$. If $\varphi(w) = 0$ then $Px = 0$ for every $x \in U$. In particular, $Px = 0$ for every $x \in V$ and

$$\psi(w) = S = P|_V = 0.$$

This shows that $\ker\varphi \subset \ker\psi$, and we are done. □

Proposition 6.8. *Let $V = \mathcal{V}(X,Y)$ and $U = \mathcal{U}(X,Y)$ for Banach operator ideals $\mathcal{V}$ and $\mathcal{U}$, and $V \subset U$. Then $\|T\|_{\mathcal{B}(U\|V)} \le \|T\|_{\widehat{\mathcal{B}}_*(U)}$ for every $T \in \widehat{\mathcal{B}}_*(U)$.*

Proof. Indeed, as $\|\cdot\|_{\widehat{\mathcal{B}}_*(U)}$ majorizes the norm $\|\cdot\|_{\mathcal{B}(U)}$ on $\widehat{\mathcal{B}}_*(U)$, we obtain from Lemma 6.7 that

$$\|T\|_{\mathcal{B}(U\|V)} = \max\left\{\|T\|_{\mathcal{B}(U)}, \|T|_V\|_{\mathcal{B}(V)}\right\} \le \|T\|_{\widehat{\mathcal{B}}_*(U)}$$

for every $T \in \widehat{\mathcal{B}}_*(U)$. □

6.3.2. Applications to semicompact multiplication operators. Let X, Y be arbitrary Banach spaces. Let $\mathcal{X} = \mathcal{B}(X,Y)$ and $\mathcal{Y} = \mathcal{N}(X,Y)$, the space of all nuclear operators $X \longrightarrow Y$. It is clear that $\mathcal{Y}$ is a Banach subspace of $\mathcal{X}$. Also, $\mathcal{X}$ and $\mathcal{Y}$ are Banach operator bimodules over the algebras $\mathcal{B}(X)$ and $\mathcal{B}(Y)$, so the algebra $\mathcal{B}(\mathcal{X}\|\mathcal{Y})$ contains the algebra $\widehat{\mathcal{B}}_*(\mathcal{X})$ of all multiplication operators

$$T = \sum_i L_{a_i} R_{b_i} \text{ with } \sum_i \|a_i\|\|b_i\| < \infty$$

where $a_i \in \mathcal{B}(Y)$, $b_i \in \mathcal{B}(X)$. Recall that a multiplication operator T is called *semicompact* if it can be written in the form

$$T = \sum_i L_{a_i} R_{t_i} + \sum_j L_{s_j} R_{b_j}, \tag{6.8}$$

where all a_i and b_j are compact operators, and

$$\sum_i \|a_i\|\|t_i\| + \sum_j \|s_j\|\|b_j\| < \infty.$$

Also, an elementary operator T is called *semifinite* if it can be written in the form (6.8) with a_i and b_j of finite rank. The algebras of all semicompact multiplication operators on $\mathcal{X}$ and all semifinite elementary operators on $\mathcal{X}$ are denoted by $\widehat{\mathcal{K}}_{\frac{1}{2}}(\mathcal{X})$ and $\mathcal{F}_{\frac{1}{2}}(\mathcal{X})$, respectively.

In particular, from above we have that $\widehat{\mathcal{K}}_{\frac{1}{2}}(\mathcal{X}) \subset \mathcal{B}(\mathcal{X}\|\mathcal{Y})$.

Theorem 6.9. *Let $\mathcal{X} = \mathcal{B}(X,Y)$ and $\mathcal{Y} = \mathcal{N}(X,Y)$. Then*

$$\widehat{\mathcal{K}}_{\frac{1}{2}}(\mathcal{X}) \subset Q_{\mathcal{B}(\mathcal{X},\mathcal{Y})}(\mathcal{B}(\mathcal{X}\|\mathcal{Y})). \tag{6.9}$$

Proof. By Corollary 4.9, we have that

$$\widehat{\mathcal{K}}_{\frac{1}{2}}(\mathcal{X}) \subset Q_{\mathcal{F}_{\frac{1}{2}}(\mathcal{X})}\left(\widehat{\mathcal{B}}_*(\mathcal{X})\right). \tag{6.10}$$

As $\|\cdot\|_{\mathcal{B}(\mathcal{X}\|\mathcal{Y})} \le \|\cdot\|_{\widehat{\mathcal{B}}_*(\mathcal{X})}$ on $\widehat{\mathcal{B}}_*(\mathcal{X})$ by Proposition 6.8, it follows that

$$Q_{\mathcal{F}_{\frac{1}{2}}(\mathcal{X})}\left(\widehat{\mathcal{B}}_*(\mathcal{X})\right) \subset Q_{\mathcal{F}_{\frac{1}{2}}(\mathcal{X})}(\mathcal{B}(\mathcal{X}\|\mathcal{Y})). \tag{6.11}$$

Since $S\mathcal{X} \subset \mathcal{Y}$ for every $S \in \mathcal{F}_{\frac{1}{2}}(\mathcal{X})$, we have that

$$\mathcal{F}_{\frac{1}{2}}(\mathcal{X}) \subset \mathcal{B}(\mathcal{X},\mathcal{Y})$$

by Theorem 6.4. Therefore, we obtain that

$$Q_{\mathcal{F}_{\frac{1}{2}}(\mathcal{X})}(\mathcal{B}(\mathcal{X}\|\mathcal{Y})) \subset Q_{\mathcal{B}(\mathcal{X},\mathcal{Y})}(\mathcal{B}(\mathcal{X}\|\mathcal{Y})), \tag{6.12}$$

and (6.9) follows from (6.10), (6.11) and (6.12). □

Now we are able to apply Theorem 6.5 to obtain the following

Theorem 6.10. *Let T be a semicompact multiplication operator on $\mathcal{B}(X,Y)$. Suppose that a closed subspace $\mathcal{Z}$ of $\mathcal{B}(X,Y)$ is invariant for T and that T is surjective on $\mathcal{Z}$. Then $\mathcal{Z}$ consists of nuclear operators, and the usual operator norm is equivalent to the nuclear norm on $\mathcal{Z}$.*

In particular, all eigenspaces of T corresponding to non-zero eigenvalues and all spectral subspaces of T corresponding to clopen subsets of $\sigma(T)$ non-containing 0 consist of nuclear operators.

The following result holds for integral semicompact operators by Proposition 4.14, Theorems 4.15 and 6.10.

Theorem 6.11. *Let $T_{a,b,s,t}$ be an integral semicompact operator on $\mathcal{X} = \mathcal{B}(X,Y)$ in the conditions of Proposition 4.14 or Theorem 4.15. Then all invariant subspaces of $T_{a,b,s,t}$ on which it is surjective consist of nuclear operators. In particular, each solution x of the equation*

$$T_{a,b,s,t}x = \lambda x$$

where $\lambda \neq 0$, is a nuclear operator.

We may apply previous results to matrix multiplication operators (see Section 4.3.2).

Corollary 6.12. *Let a matrix $(T_{pq})_{p,q=1}^n$ consist of semicompact multiplication operators and let T be the matrix multiplication operator defined by this matrix. Then the spectral subspaces of T that correspond to clopen subsets of $\sigma(T)$ non-containing 0, consist of n-tuples of nuclear operators.*

Proof. Let $\mathcal{X} = \mathcal{B}(X,Y)^{(n)}$, the direct sum of n copies of $\mathcal{B}(X,Y)$, $\mathcal{Y} = \mathcal{N}(X,Y)^{(n)}$ and $U = \mathcal{B}(X,Y)$. Then it is easy to see that

$$\mathbb{M}_n\left(\widehat{\mathcal{K}}_{\frac{1}{2}}(U)\right) \subset \mathbb{M}_n\left(\widehat{\mathcal{B}}_*(U)\right) \subset \mathcal{B}(\mathcal{X}\|\mathcal{Y}) \text{ and } \mathbb{M}_n\left(\mathcal{F}_{\frac{1}{2}}(U)\right) \subset \mathcal{B}(\mathcal{X},\mathcal{Y}).$$

Now a similar argument as in Theorem 6.9 shows that

$$\mathbb{M}_n(\widehat{\mathcal{K}}_{\frac{1}{2}}(U)) \subset Q_{\mathcal{B}(\mathcal{X},\mathcal{Y})}(\mathcal{B}(\mathcal{X}||\mathcal{Y})),$$

and it remains to apply Theorem 6.5. □

6.4. Semicompact elementary operators

Let X, Y be Banach spaces. Assume now that T is an elementary operator on $\mathcal{B}(X,Y)$:

$$T = \sum_{i=1}^{n} L_{a_i} R_{x_i} + \sum_{j=1}^{k} L_{y_j} R_{b_j},$$

where all $x_i \in \mathcal{B}(X)$, $y_j \in \mathcal{B}(Y)$, $a_i \in \mathcal{K}(Y)$, $b_j \in \mathcal{K}(X)$. According to our terminology, T is a *semicompact elementary operator.*

Our aim is to show that the statement of Theorem 6.10 in this case can be considerably strengthened: invariant subspaces on which T is surjective are contained in the component $\mathfrak{J}(X,Y)$ of each quasi-Banach operator ideal $\mathfrak{J}$. In this situation the approach based on the tensor products of Banach algebras and the tensor spectral radius theory is not directly applicable and for the proof that some power of T is close (in a proper sense) to a semifinite elementary operator, we use the arguments based on the analysis of triangularizable sets of compact operators.

6.4.1. Quasi-Banach operator ideals. Recall that *a quasinorm* on a linear space $\mathcal{L}$ is a map $\|\cdot\|_{\mathcal{L}} : \mathcal{L} \to \mathbb{R}^+$ satisfying the conditions

$$\begin{aligned} \|x+y\|_{\mathcal{L}} &\le t_{\mathcal{L}}(\|x\|_{\mathcal{L}} + \|y\|_{\mathcal{L}}) \text{ for all } x, y \in \mathcal{L} \text{ and some } t_{\mathcal{L}} \ge 1, \\ \|\lambda x\|_{\mathcal{L}} &= |\lambda| \, \|x\|_{\mathcal{L}} \text{ for all } \lambda \in \mathbb{C}, x \in \mathcal{L}, \text{ and} \\ \|x\|_{\mathcal{L}} &= 0 \text{ iff } x = 0. \end{aligned} \tag{6.13}$$

By [17, Page 162], each quasinorm generates a linear (metrizable) Hausdorff topology on $\mathcal{L}$. We say that $\mathcal{L}$ is *complete under the quasinorm* if it is complete in this topology.

Furthermore, a *quasi-Banach operator ideal* $\mathfrak{J}$ (see [20]) consists of components $\mathfrak{J}(X,Y) \subset \mathcal{B}(X,Y)$ complete under a quasinorm $\|\cdot\|_{\mathfrak{J}(X,Y)} = \|\cdot\|_{\mathfrak{J}}$, where X and Y run over Banach spaces, and satisfying the following conditions

1) $t_{\mathfrak{J}(X,Y)} = t_{\mathfrak{J}}$ for some $t_{\mathfrak{J}} \ge 1$ and all Banach spaces X and Y, where $t_{\mathfrak{J}(X,Y)}$ is the constant $t_{\mathcal{L}}$ in (6.13) for $\mathcal{L} = \mathfrak{J}(X,Y)$.

2) $\|axb\|_{\mathfrak{J}} \le \|a\| \, \|x\|_{\mathfrak{J}} \, \|b\|$ for all $x \in \mathfrak{J}(X,Y)$, $a \in \mathcal{B}(Y,Z)$, $b \in \mathcal{B}(W,X)$, where Z and W run over Banach spaces,

3) $\|x\|_{\mathfrak{J}} = \|x\|$ for each operator x of rank one.

By [20, Theorem 6.2.5], each quasi-Banach ideal $\mathfrak{J}$ has an equivalent quasinorm $|\cdot|_{\mathfrak{J}}$ with the property that there is a number p such that $0 < p \le 1$ and

$$|x+y|_{\mathfrak{J}}^p \le |x|_{\mathfrak{J}}^p + |y|_{\mathfrak{J}}^p \tag{6.14}$$

for every $x, y \in \mathfrak{J}(X,Y)$ and for all Banach spaces X, Y (one can take p as a number satisfying $(2t)^p = 2$ for $t \geq t_{\mathfrak{J}}$). We assume that a quasinorm in consideration satisfies this condition, and write $\|\cdot\|_p$ or $\|\cdot\|_{p,\mathfrak{J}}$ instead of $|\cdot|_{\mathfrak{J}}$. In this case we say that $\mathfrak{J}$ is a *p-Banach operator ideal.* It should be noted that the topology of $\mathfrak{J}(X,Y)$ is given by the metric $\mathrm{d}(x,y) = \|x-y\|_p^p$. In the same way we denote the corresponding quasinorm on bounded operators T on $\mathfrak{J}(X,Y)$:

$$\|T\|_p = \|T\|_{p,\mathfrak{J}} = \inf\{t > 0 : \|Tx\|_p \leq t\|x\|_p \text{ for all } x \in \mathfrak{J}(X,Y)\}.$$

Lemma 6.13. *Let $\mathfrak{J}$ be a p-Banach operator ideal, and let T be an elementary operator on $\mathcal{B}(X,Y)$, $Tx = \sum_{i=1}^n a_i x b_i$ for every $x \in \mathcal{B}(X,Y)$, where $a_i \in \mathcal{B}(Y)$, $b_i \in \mathcal{B}(X)$. Then T is bounded on $\mathfrak{J}(X,Y)$ and*

$$\|Tx\|_p \leq n^{\frac{1-p}{p}} \left(\sum_{i=1}^n \|a_i\|\|b_i\|\right) \|x\|_p$$

for all $x \in \mathfrak{J}(X,Y)$.

Proof. It follows from (6.14) under $|\cdot|_{\mathfrak{J}} = \|\cdot\|_p$ that

$$\|Tx\|_p^p \leq \sum_{i=1}^n \|a_i\|^p \|b_i\|^p \|x\|_p^p$$

for all $x \in \mathfrak{J}(X,Y)$.

Since the function $f(t) = t^p$ is concave for $t \geq 0$ and $0 < p \leq 1$, we obtain that

$$\sum_{i=1}^n t_i^p \leq n^{1-p} \left(\sum_{i=1}^n t_i\right)^p. \tag{6.15}$$

Applying this to $t_i = \|a_i\|\|b_i\|$, we get that

$$\|Tx\|_p^p \leq n^{1-p} \left(\sum_{i=1}^n \|a_i\|\|b_i\|\right)^p \|x\|_p^p$$

which gives what we need. □

In a short form the statement of the previous lemma can be written as follows:

$$\text{If } T = \sum_{i=1}^n L_{a_i} R_{b_i} \text{ then } \|T\|_p \leq n^{\frac{1-p}{p}} \sum_{i=1}^n \|a_i\|\|b_i\|.$$

Similarly we obtain the following

Lemma 6.14. *Let a be a finite rank operator in $\mathcal{B}(X,Y)$. Then*

$$\|a\|_p \leq n^{\frac{1-p}{p}} \|a\|_{\mathcal{N}(X,Y)},$$

where n is the rank of a and $\|\cdot\|_p$ is the p-norm of a p-Banach operator ideal $\mathfrak{J}$.

Proof. It is easy to check that for any $\varepsilon > 0$ there are rank one operators a_i such that

$$a = \sum_{i=1}^{n} a_i,$$

where n is the rank of a, and

$$\sum_{i=1}^{n} \|a_i\| \le \|a\|_{\mathcal{N}(X,Y)} + \varepsilon.$$

On the other hand,

$$\|a\|_p^p \le \sum_{i=1}^{n} \|a_i\|_p^p = \sum_{i=1}^{n} \|a_i\|^p \le n^{1-p} \left(\sum_{i=1}^{n} \|a_i\| \right)^p \le n^{1-p} \left(\|a\|_{\mathcal{N}(X,Y)} + \varepsilon \right)^p$$

by (6.15). As ε is arbitrary, we obtain that $\|a\|_p \le n^{\frac{1-p}{p}} \|a\|_{\mathcal{N}(X,Y)}$. □

6.4.2. Quasinilpotence of semicompact elementary operators modulo semifinite ones with respect to a quasinorm. If W is a closed subspace of a Banach space X then for each $x \in X$, we will write x/W instead of $x + W$ for the corresponding element of X/W. If moreover a is an operator on X leaving W invariant then we denote by $a|_W$ and $a|_{X/W}$ its restriction to W and, respectively, the operator induced by a in X/W.

Lemma 6.15. *Let W be a closed subspace of a Banach space X, and let a, b be operators on X which preserve W invariant. Then*

$$\|ab\| \le 2\|a|_W\| \|b\| + \|a\| \|b|_{X/W}\|. \tag{6.16}$$

Proof. For any $x \in X$ and $\varepsilon > 0$, choose $y \in W$ with $\|bx - y\| \le \| (bx)/W\|_{X/W} + \varepsilon$. Then

$$\|y\| \le \|bx\| + \| (bx)/W\|_{X/W} + \varepsilon \le 2\|bx\| + \varepsilon,$$

whence we obtain that

$$\begin{aligned} \|abx\| &\le \|ay\| + \|a\| \|bx - y\| \le \|a|_W\| \|y\| + \|a\|(\| (bx)/W\|_{X/W} + \varepsilon) \\ &\le \|a|_W\|(2\|bx\| + \varepsilon) + \|a\|(\|b|_{X/W}\| \|x/W\|_{X/W} + \varepsilon). \end{aligned}$$

Since $\|x/W\|_{X/W} \le \|x\|$ and ε is arbitrary, we obtain that

$$\|abx\| \le (2\|a|_W\| \|b\| + \|a\| \|b|_{X/W}\|)\|x\|$$

which is what we need. □

Lemma 6.16. *Let $0 = X_0 \subset X_1 \subset X_2 \subset \cdots \subset X_k \subset X$ be a chain of closed subspaces in a Banach space X. Let $m \ge k$ and let $a_1, \dots, a_m \in \mathcal{B}(X)$ preserve all X_j invariant: $a_i X_j \subset X_j$. If $\|a_i\| \le \alpha$ for all i, and $\|a_i|_{X_j/X_{j-1}}\| \le \beta$ for all i, j, then*

$$\|a_1 a_2 \cdots a_m\| \le 2^m C_m^k \alpha^k \beta^{m-k}.$$

Proof. We use induction in m, k. For the base of the induction, note that the statement is evidently true for $k = 0$ and for $m = k$, and Lemma 6.15 establishes it for $k = 1, m = 2$. Now assuming that the statement holds for $(m-1, k-1)$ and $(m-1, k)$ we prove that it holds for (m, k). Indeed, setting $W = X_1$, $a = a_1$, $b = a_2 \cdots a_m$ in the notation of Lemma 6.15, we obtain from (6.16) that

$$\|a_1 a_2 \cdots a_m\| \le 2\|a_1|_{X_1}\|\|a_2 \cdots a_m\| + \|a_1\|\|a_2 \cdots a_m|_{X/X_1}\|.$$

By the induction assumption, we have that

$$\|a_2 \cdots a_m\| \le 2^{m-1} C_{m-1}^k \alpha^k \beta^{m-1-k}.$$

Furthermore, the operators $a_i|_{X/X_1}$ preserve the chain $\{X_i/X_1 : i \le k\}$ which consists of $k-1$ non-trivial elements. Hence again by the induction assumption, we obtain that

$$\|a_2 \cdots a_m|_{X/X_1}\| \le 2^{m-1} C_{m-1}^{k-1} \alpha^{k-1} \beta^{m-k}.$$

Therefore

$$\begin{aligned}\|a_1 a_2 \cdots a_m\| &\le 2\beta 2^{m-1} C_{m-1}^k \alpha^k \beta^{m-1-k} + \alpha 2^{m-1} C_{m-1}^{k-1} \alpha^{k-1} \beta^{m-k} \\ &\le 2^m \alpha^k \beta^{m-k} (C_{m-1}^k + C_{m-1}^{k-1}) = 2^m C_m^k \alpha^k \beta^{m-k}.\end{aligned}$$

□

Lemma 6.17. *Let K be a finite set of compact operators in the radical of an operator algebra $A \subset \mathcal{B}(X)$, and let F be a bounded subset of A. Let $\lambda \in (0,1)$. For each m, let $E_{K,F}^\lambda(m)$ denote the set of all products $b_1 \dots b_m$ of elements in $K \cup F$ in which the number of those b_i that belong to K is greater than or equal to λm. Then $\|E_{K,F}^\lambda(m)\|^{1/m} \to 0$ for $m \to \infty$.*

Proof. Without loss of generality, one may assume that $\|K \cup F\| = 1$. By [24] (see also [25]), for each $\varepsilon > 0$ there is a finite chain $0 \subset X_1 \subset \cdots \subset X_k \subset X$ of invariant subspaces for A such that $\|b|_{X_j/X_{j-1}}\| \le \varepsilon$ for all $b \in K$ and all $j \le k$. It follows that

$$\|c|_{X_j/X_{j-1}}\| \le \varepsilon$$

if $c = b_1 \dots b_p a$, where $b_i \in F$ and $a \in K$.

Each product $b_1 \dots b_m \in E_{K,F}^\lambda(m)$ can be written in the form $c_1 c_2 \dots c_l$, where all of c_i are as above and $l \ge \lambda m$. Applying the result of Lemma 6.16, we obtain that

$$\|b_1 \dots b_m\| \le 2^l C_l^k \varepsilon^l \le 2^m m^k \varepsilon^{\lambda m}.$$

Thus

$$\|E_{K,F}^\lambda(m)\|^{1/m} \le 2 m^{k/m} \varepsilon^\lambda \le 3\varepsilon^\lambda$$

for sufficiently big m. Since ε is arbitrary, we are done. □

A subset M of a Banach algebra A is called *bicompact* if $L_a R_b$ is a compact operator for every $a, b \in M$.

Proposition 6.18. *Let M be a finite bicompact subset of A in the radical of a Banach algebra A, and let N be a bounded subset of A. For each m, let $H(m)$ denote the set of all products $x_1 \cdots x_m$ of elements in $M \cup N$ in which the number of those x_i that belong to M is greater than or equal to $m/2$. Then $\|H(m)\|^{1/m} \to 0$ under $m \to \infty$.*

Proof. We may assume that A is unital and 1 is the unit of A. Let K be the set of all operators $L_a R_b$ on A, where $a, b \in M$. Let $F = \{L_a : a \in N\} \cup \{R_a : a \in N\}$.

We claim that every product w in $H(m)$ can be written as $T(1)$, where T is a product of operators in which the number of operators in K is greater than or equal to $[m/4]$ (and the number of operators in F is less than or equal to $m/2+1$). Indeed, we do as follows. Represent w as the product of w_1 and w_2 in which of each the number of those x_i that belong to M is greater than or equal to $[m/4]$. Let $a_0 = 1$, $w_1 = w_3 x_i v_1$ and $w_2 = v_2 x_j w_4$ for some $x_i, x_j \in M$, where v_1 and v_2 do not contain any elements from M as a factor. Then $w = w_3 a_1 w_4$, where $a_1 = S_1 L_{v_1} R_{v_2}(a_0)$ and $S_1 = L_{x_i} R_{x_j} \in K$. Arguing by induction, we obtain that $w = w_{2k+1} a_k w_{2k+2}$, where $a_k = S_k P_k(a_{k-1})$, $S_k \in K$ and P_k is a product of operators in F, for $k \le [m/4]$. So we obtain the required representation w as $T(1)$ for some $k \ge [m/4]$.

Now it follows in the notation of Lemma 6.17 that

$$\|H(m)\| \le \left\| E_{K,F}^{1/3}\left(\frac{2m}{3}\right) \right\|$$

for sufficiently big m, and it remains to apply Lemma 6.17. □

Theorem 6.19. *Let $T = \sum_{i=1}^{n} L_{a_i} R_{x_i} + \sum_{j=1}^{k} L_{y_j} R_{b_j}$ be a semicompact elementary operator on $\mathcal{B}(X,Y)$, where all $x_i \in \mathcal{B}(X)$, $y_j \in \mathcal{B}(Y)$, $a_i \in \mathcal{K}(Y)$, $b_j \in \mathcal{K}(X)$. Then for any $\varepsilon > 0$, there is $m \in \mathbb{N}$ and an operator $S = \sum_{i=1}^{(n+k)^m} L_{c_i} R_{d_i}$ such that*

$$\|T^m - S\|_p < \varepsilon^m$$

and c_i or d_i is of finite rank for each i, where $\|\cdot\|_p$ is the p-norm of a p-Banach operator ideal $\mathfrak{J}$.

Proof. A required decomposition of the operator T^m into the sum of $(n+k)^m$ summands can be written as

$$T^m = T_1 + T_2,$$

where in T_1 we gather those summands where the number of factors L_{a_i} is more than the number of factors L_{y_j} (hence their number $\ge m/2$), while summands in T_2 have more factors R_{b_j} than factors R_{x_i}.

Let $A = \mathcal{B}(Y)/\mathcal{A}(Y)$ and $q : \mathcal{B}(Y) \longrightarrow A$ be the standard epimorphism. Let $M = \{q(a_1), \ldots, q(a_n)\}$ and $N = \{q(y_1), \ldots, q(y_k)\}$. Then M is a bicompact subset of A in the radical of A. Writing T_1 as $\sum L_{w_i} R_{z_i}$, where $L_{w_i} R_{z_i}$ are the above summands of T_1 in the decomposition of T^m, we note that the corresponding family $H(m)$ (see the above lemma) consists of all products of elements which are

q-images of left coefficients w_i of summands in T_1. By Lemma 6.18, there is m such that

$$\|H(m)\| < \varepsilon^m.$$

This means that for every w_i there is a finite rank operator r_i with

$$\|r_i - w_i\| < \varepsilon^m.$$

Then, setting $S_1 = \sum L_{r_s} R_{z_s}$, we obtain a semifinite multiplication operator such that $S_1 - T_1$ can be represented in the form $\sum L_{v_i} R_{z_i}$ with

$$\sum_i \|v_i\| \|z_i\| < (n+k)^m \varepsilon^m.$$

For brevity, we rewrite this in the form

$$\|S_1 - T_1\|_* < (n+k)^m \varepsilon^m.$$

Similarly, we find a semifinite operator $S_2 = \sum L_{e_s} R_{f_s}$, where all f_s are finite rank operators, with

$$\|S_2 - T_2\|_* < (n+k)^m \varepsilon^m.$$

Hence setting $S = S_1 + S_2$ we obtain that S is semifinite and

$$\|T - S\|_* < 2(n+k)^m \varepsilon^m.$$

As the number of elementary summands (of length one) in S is $(n+k)^m$ by our choice then we obtain that

$$\|T^m - S\|_p < ((n+k)^m)^{\frac{1-p}{p}} 2(n+k)^m \varepsilon^m.$$

by Lemma 6.13. Changing ε by $\gamma\varepsilon$ for sufficiently small γ, we obtain the required inequality. □

6.4.3. Spectral subspaces of semicompact elementary operators. Now we are able to prove the following

Theorem 6.20. *Let $T = \sum_{i=1}^n L_{a_i} R_{x_i} + \sum_{j=1}^k L_{y_j} R_{b_j}$ be a semicompact elementary operator on $\mathcal{B}(X,Y)$, where all $x_i \in \mathcal{B}(X)$, $y_j \in \mathcal{B}(Y)$, $a_i \in \mathcal{K}(Y)$, $b_j \in \mathcal{K}(X)$. Suppose that $T\mathcal{Z} = \mathcal{Z}$ for a closed subspace $\mathcal{Z}$ of $\mathcal{B}(X,Y)$. Then $\mathcal{Z}$ is contained in $\mathfrak{J}(X,Y)$ for any quasi-Banach operator ideal $\mathfrak{J}$.*

Proof. One may suppose that $\mathfrak{J}$ is a p-Banach operator ideal with p-norm $\|\cdot\|_p$ for $0 < p \le 1$.

By the Open Mapping Theorem, there is $t > 0$ such that for each $z \in \mathcal{Z}$ there is $w \in \mathcal{Z}$ with $Tw = z$ and $\|w\| \le t\|z\|$. Take $\varepsilon > 0$ such that $\varepsilon < t^{-1}$, and choose m and S as in Theorem 6.19. Setting $P = T^m - S$, we have that $\|P\|_p \le \varepsilon^m$.

On the other hand, S maps each operator from $\mathcal{B}(X,Y)$ into an operator of rank $\le d$, where d is the sum of the ranks of finite rank coefficients of S. As $\mathfrak{J}$ contains all finite rank operators, we have that $Sx \in \mathfrak{J}(X,Y)$ and

$$\|Sx\|_p \le d^{\frac{1-p}{p}} \|Sx\|_{\mathcal{N}(X,Y)}$$

by Lemma 6.14, for every $x \in \mathcal{B}(X,Y)$. It follows from Theorem 6.4 that S is bounded as an operator $\mathcal{B}(X,Y) \longrightarrow \mathcal{N}(X,Y)$. Let s be its norm as such an operator. Then

$$\|Sx\|_{\mathcal{N}(X,Y)} \leq s\,\|x\|$$

for every $x \in \mathcal{B}(X,Y)$. As a result, we obtain that

$$\|Sx\|_p \leq d^{\frac{1-p}{p}} s\,\|x\| \tag{6.17}$$

for all $x \in \mathcal{B}(X,Y)$.

Now we may argue as in the proof of Theorem 6.5. Then, as we saw, each $z \in \mathcal{Z}$ can be expressed as in (6.5):

$$z = \sum_{j=0}^{\infty} P^j S z_{j+1}$$

with the estimation

$$\|z_j\| \leq t^{mj}\,\|z\| \tag{6.18}$$

for every j (see (6.4) in Theorem 6.5). As all $P^j S z_{j+1} \in \mathfrak{J}(X,Y)$, we may estimate their p-norms $\|P^j S z_{j+1}\|_p$ as follows. We have that

$$\|P^j S z_{j+1}\|_p \leq \|P\|_p^j \|S z_{j+1}\|_p$$

and that

$$\|S z_{j+1}\|_p \leq d^{\frac{1-p}{p}} s \|z_{j+1}\| \leq d^{\frac{1-p}{p}} s t^{m(j+1)} \|z\|$$

by (6.17) and (6.18). By our choice, we have that $\|P\|_p \leq \varepsilon^m$. So we obtain that

$$\|P^j S z_{j+1}\|_p \leq \varepsilon^{mj} d^{\frac{1-p}{p}} s t^{m(j+1)} \|z\| = \left(d^{\frac{1-p}{p}} s t^m \|z\|\right) (\varepsilon t)^{mj},$$

whence

$$\sum_{j=0}^{\infty} \|P^j S z_{j+1}\|_p^p \leq \left(d^{\frac{1-p}{p}} s t^m \|z\|\right)^p \sum_{j=0}^{\infty} ((\varepsilon t)^{mp})^j < \infty$$

because of $(\varepsilon t)^{mp} < 1$. As $\mathfrak{J}(X,Y)$ is complete under $\|\cdot\|_p$, the convergence of this series implies that $z \in \mathfrak{J}(X,Y)$. □

As a consequence, we obtain the following

Corollary 6.21. *Let T be a semicompact elementary operator on $\mathcal{B}(X,Y)$. Then all eigenspaces of T corresponding to non-zero eigenvalues and all spectral subspaces of T corresponding to clopen subsets of $\sigma(T)$ non-containing* 0 *are contained in $\mathfrak{J}(X,Y)$ for any quasi-Banach operator ideal $\mathfrak{J}$.*

References

[1] J.C. Alexander, Compact Banach algebras, *Proc. London Math. Soc.* (3) **18** (1968), 1–18.

[2] B. Aupetit, *Propriétés spectrales des algèbres de Banach*, Lect. Notes Math., 735, Springer-Verlag, Berlin, 1979.

[3] B. Aupetit, M. Mathieu, The continuity of Lie homomorphisms, *Studia Math.* **138** (2) (2000), 193–199.

[4] B.A. Barnes, Banach algebras which are ideals of a Banach algebra, *Pacific J. Math.* **38**, 1 (1971), 1–7.

[5] B.A. Barnes, Density theorems for algebras of operators and annihilator Banach algebras, *Mich. Math. J.* **19** (1972), 149–155.

[6] J. Bergh and J. Löfström, *Interpolation spaces, An introduction*, Springer-Verlag, Berlin, 1976.

[7] M.A. Berger and Y. Wang, Bounded semigroups of matrices, *Linear algebra Appl.* 166 (1992), 21–27.

[8] F.F. Bonsall, Operators that act compactly on an algebra of operators, *Bull. London Math. Soc.* **1** (2) (1969), 163–170.

[9] F.F. Bonsall, J. Duncan, *Complete normed algebras*, Springer-Verlag, Berlin, Heidelberg, New York, 1973.

[10] N. Bourbaki, *Elements of mathematics, Algebra* 1, *Chapters* 1–3, Springer-Verlag, Berlin, 1998.

[11] R.E. Curto, Spectral theory of elementary operators, in "*Elementary operators and applications*", Editor M. Mathieu, World Sci. Publ., Singapore, New Jersey, London, 1992, 3–54.

[12] A. Defant, K. Floret, *Tensor norms and operator ideals*, Elsevier, Amsterdam, 1993.

[13] P.G. Dixon, Topologically irreducible representations and radicals in Banach algebras, *Proc. London Math. Soc.* (3) **74** (1997), 174–200.

[14] C.-K. Fong, H. Radjavi, On ideals and Lie ideals of compact operators, *Math. Ann.* **262** (1983), 23–28.

[15] E. Hille, R. Phillips, *Functional analysis and semigroups*, AMS, Providence, 1957.

[16] M. Kennedy, V.S. Shulman, Yu.V. Turovskii, Invariant subspaces of subgraded Lie algebras of compact operators, *Integr. Eq. Oper. Theory* **63** (2009), 47–93.

[17] G. Köthe, *Topological vector spaces I*, Springer-Verlag, New York, 1969.

[18] V.I. Lomonosov, Invariant subspaces for operators commuting with compact operators, *Funct. Anal. Appl.* **7** (1973), 213–214.

[19] C.J. Read, Quasinilpotent operators and the invariant subspace problem, *J. London Math. Soc.* (2) **56** (1997), 595–606.

[20] A. Pietsch, *Operator ideals*, VEB Deutscher Verlag der Wissenschaften, Berlin, 1978.

[21] G.-C. Rota and W.G. Strang, A note on the joint spectral radius, *Indag. Math.* **22** (1960) 379–381.

[22] W. Rudin, *Functional analysis*, McGraw-Hill, New York, 1991.

[23] A.F. Ruston, *Fredholm theory in Banach spaces*, Cambridge Univ. Press, Cambridge, 1986.

[24] V.S. Shulman, On invariant subspaces of Volterra operators, *Functional Anal. i Prilozen.* **18** (2) (1984), 84–85 (in Russian).

[25] V.S. Shulman, Yu.V. Turovskii, Joint spectral radius, operator semigroups and a problem of W. Wojtyński, *J. Funct. Anal.* **177** (2000), 383–441.

[26] V.S. Shulman, Yu.V. Turovskii, Radicals in Banach algebras and some problems in the theory of radical Banach algebras, *Funct. Anal. Appl.* **35** (2001), 312–314.

[27] V.S. Shulman, Yu.V. Turovskii, Formulae of joint spectral radii for sets of operators, *Studia Math.* **149** (2002), 23–37.

[28] V.S. Shulman, Yu.V. Turovskii, Invariant subspaces of operator Lie algebras and Lie algebras with compact adjoint action, *J. Funct. Anal.* **223** (2005), 425–508.

[29] V.S. Shulman, Yu.V. Turovskii, Topological radicals, I. Basic properties, tensor products and joint quasinilpotence, *Banach Center Publ.*, 67 (2005), 293–333.

[30] V.S. Shulman, Yu.V. Turovskii, Application of topological radicals to calculation of joint spectral radii, *preprint*: arXiv:0805.0209 (2 May 2008).

[31] Yu.V. Turovskii, Volterra semigroups have invariant subspaces, *J. Funct. Anal.* **162** (2) (1999), 313–323.

[32] K. Vala, On compact sets of compact operators, *Ann. Acad. Sci. Fenn.* Ser. A I **351** (1964), 1–8.

[33] E. Vesentini, On the subharmonicity of the spectral radius, *Boll. Un. Mat. Ital.* **4** (1968), 427–429.

[34] G. Willis, Compact approximation property does not imply approximation property, *Studia Math.* **103** (1992), 99–108.

[35] W. Wojtynski, A note on compact Banach-Lie algebras of Volterra type, Bull. Acad. Polon. Sci., Ser. sci. math., astr. et phys. 26 (1978), 2, 105–107.

[36] W. Wojtynski, On the existence of closed two-sided ideals in radical Banach algebras with compact elements, Bull. Acad. Polon. Sci., Ser. sci. math., astr. et phys. 26 (1978), 2, 109–113.

[37] W. Wojtynski, Quasinilpotent Banach-Lie algebras are Baker-Campbell-Hausdorff, *J. Funct. Anal.* **153** (1998), 405–413.

[38] E.I. Zelmanov, On Engel Lie algebras, *DAN USSR*, **292** (1987) no. 2, 265–268 (in Russian).

V.S. Shulman
Department of Mathematics,
Vologda State Technical University,
15 Lenina str.
Vologda 160000, Russian Federation
e-mail: shulman_v@yahoo.com

Yu.V. Turovskii
Institute of Mathematics and Mechanics,
National Academy of Sciences of Azerbaijan,
9 F. Agayev Street
Baku AZ1141, Azerbaijan
e-mail: yuri.turovskii@gmail.com

An Elementary Approach to Elementary Operators on $\mathcal{B}(H)$

V.S. Shulman and L. Turowska

Abstract. We review some properties of elementary operators on the space of bounded operators on a Hilbert space H which they share with operators on H.

Mathematics Subject Classification (2000). Primary 47B47; Secondary 47A62, 47L05, 47B10 .

Keywords. Elementary operator, normal, derivation.

1. Introduction

In 1975 J. Anderson and F. Foias wrote a paper [2] whose title could be used (with modifications) in many later papers on elementary operators. In the present paper we discuss properties which operators on a Hilbert space H share with elementary operators on $\mathcal{B}(H)$. The reason of the search of common properties is that an elementary operator $\Delta = \sum_{k=1}^{n} L_{a_k} R_{b_k}$ has a formal adjoint $\tilde{\Delta} = \sum_{k=1}^{n} L_{a_k^*} R_{b_k^*}$ which turns into a proper adjoint if restricted to the ideal $\mathcal{S}_2$ of Hilbert-Schmidt operators, and so it is natural to expect that their adjointness on $\mathcal{B}(H)$ is not absolutely formal. After discussion of the general case we come to normal elementary operators (that is those whose coefficient families are commutative and consists of normal operators) and study which properties of normal operators on Hilbert space they share.

Our paper is a kind of review but some formulations of the results seem to be new. Several statements of the paper can be extended to a more wide class of multiplication operators, but we prefer to restrict ourselves to elementary operators which allows us to avoid more complicated technique of Varopoulos algebras, Haagerup tensor products, spectral synthesis and so on. Moreover we hope that such approach makes the subject really elementary and gives the possibility to present the results with ideas of their proofs in a text of reasonable length. The

The second author was supported by the Swedish Research Council.

R.E. Curto and M. Mathieu (eds.), *Elementary Operators and Their Applications*, Operator Theory: Advances and Applications 212, DOI 10.1007/978-3-0348-0037-2_6, © Springer Basel AG 2011

reader can find transparent proofs and various extensions of the main part of our results in [19, 22, 23].

1.1. Notations

Let H be a Hilbert space, $\mathcal{B}(H)$ be the space of bounded linear operators on H. We denote by $\mathcal{S}_p$, $1 \le p < \infty$, a Schatten-von Neumann ideal and write $||\cdot||_p$ for the corresponding norm and let $\mathcal{S}_\infty$ to denote the space of compact operators.

Let us say that an operator (or a family of operators) is *multicyclic* or *has finite multicyclicity* if there is a finite set of vectors which is not contained in a proper closed invariant subspace.

For a subset M of a metric space we say that its Hausdorff dimension does not exceed a number $r > 0$ if there exists $C > 0$ such that for $\epsilon > 0$ there is a covering $\mathcal{B} = \{\beta_j\}$ of M by pairwise disjoint Borel sets with $\operatorname{diam}\beta_j < \epsilon$ and $|\mathcal{B}|_r := (\sum_j (\operatorname{diam}\beta_j)^r \le C$.

If $\mathbb{A} = (A_1, \dots, A_n)$ is a family of commuting normal operators then by $\sigma(\mathbb{A})$ we denote the joint spectrum, and by $E_{\mathbb{A}}(\cdot)$ the spectral measure of $\mathbb{A}$.

We say that the *essential dimension* of $\mathbb{A}$ does not exceed $r > 0$ (and write ess-dim $\mathbb{A} \le r$) if there is a subset D of $\sigma(\mathbb{A})$ such that $E_{\mathbb{A}}(\sigma(\mathbb{A}) \setminus D) = 0$ and $\dim(D) \le r$.

2. Approximate inverse intertwinings

The proofs of many further statements are based on several results of very general nature which we gather in the present section. The reader can consider them as exercises in functional analysis, (s)he can also find their solutions in Section 6 of [22].

Let $\mathfrak{X}$ and $\mathfrak{Y}$ be topological vector spaces, $\Phi : \mathfrak{X} \to \mathfrak{Y}$ a continuous imbedding with dense range, and let S and T be operators acting in $\mathfrak{X}$ and $\mathfrak{Y}$, respectively, intertwined by the mapping Φ: $T\Phi = \Phi S$. We write in this case that we are given an intertwining triple (or just an intertwining) (Φ, S, T).

A net of linear mappings $F_\alpha : \mathfrak{Y} \to \mathfrak{X}$ is called an *approximate inverse intertwining* (AII) for the intertwining (Φ, S, T) if

(a) $F_\alpha \Phi \to 1_{\mathfrak{X}}$,
(b) $\Phi F_\alpha \to 1_{\mathfrak{Y}}$ and
(c) $F_\alpha T - S F_\alpha \to 0_{\mathfrak{X}}$

in the topology of simple convergence.

Denote by Φ^{-1} the full inverse image under the mapping Φ: $\Phi^{-1}(M) = \{x \in \mathfrak{X} \mid \Phi(x) \in M\}$ for any $M \subset \mathfrak{Y}$ (non-necessarily $M \subset \Phi(\mathfrak{X})$). As usual the range of a map X is denoted by $\operatorname{im} X$.

Theorem 2.1. *If the intertwinings (Φ, S_i, T_i), $1 \le i \le n$, have a common AII, then*

$$\Phi^{-1}\Big(\sum_i \operatorname{im} T_i\Big) \subset \overline{\sum_i \operatorname{im} S_i}.$$

Let $\mathcal{H}$ be a Hilbert space equipped with the weak operator topology.

Corollary 2.2. *If $\mathfrak{X} = \mathcal{H}$ and (Φ, S, T) has an AII, then*

$$\Phi(\ker S^*) \cap \operatorname{im} T = \{0\}.$$

Theorem 2.3. *Let Φ intertwine pairs S_i, T_i ($i = 1, 2$). Suppose that $\mathfrak{X}$ is a Banach space equipped with a weak topology, and $||S_2 x|| \le ||S_1 x||$ for any $x \in \mathfrak{X}$. If (Φ, S_1, T_1) has AII then*

$$T_1^{-1}(\operatorname{im}\Phi) \subset T_2^{-1}(\operatorname{im}\Phi)$$

and

$$||\Phi^{-1} T_2 y|| \le ||\Phi^{-1} T_1 y|| \tag{2.1}$$

for any $y \in T_1^{-1}(\operatorname{im}\Phi)$.

The following result is an immediate consequence of Theorem 2.3.

Corollary 2.4. *Let $\mathfrak{X} = \mathcal{H}$. Suppose that S is a normal operator on $\mathcal{H}$ and that the intertwinings (Φ, S, T_1), (Φ, S^*, T_2) have AII's (not necessarily coinciding). Then $T_1^{-1}(\operatorname{im}\Phi) = T_2^{-1}(\operatorname{im}\Phi)$ and*

$$||\Phi^{-1} T_2 y|| = ||\Phi^{-1} T_1 y||$$

for any $y \in T_1^{-1}(\operatorname{im}\Phi) = T_2^{-1}(\operatorname{im}\Phi)$. In particular, $\ker T_1 = \ker T_2$.

Let (Φ, S, T) be an intertwining. If $\mathfrak{X}$ is a dual Banach space with the weak-* topology (for example if $\mathfrak{X} = \mathcal{H}$) then to obtain an AII it suffices to construct a net of operators $F_\alpha : \mathfrak{Y} \to \mathfrak{X}$ which satisfies a weakened version of (AII)-conditions:

(a′) $F_\alpha \Phi x$ is bounded for each $x \in \mathfrak{X}$,
(b′) $\Phi F_\alpha \to 1_{\mathfrak{Y}}$ and
(c′) $(F_\alpha T - S F_\alpha) y$ is bounded for each $y \in Y$.

In general a net satisfying ((a′), (b′), (c′)) is called *an approximate inner semiintertwining* for (Φ, S, T) (AIS, for short).

Denote by $\mathfrak{X}^*$ the space of continuous antilinear functionals on $\mathfrak{X}$, endowed with the weak-* topology (in particular, $\mathcal{H}^* = \mathcal{H}$). The adjoint operators (on $\mathcal{X}^*$ or between $\mathcal{X}^*$ and $\mathcal{Y}^*$) are defined in the usual way. In particular, the adjoint of an operator on $\mathcal{H}$ has the usual meaning.

It is not difficult to see that if $\{F_\alpha\}$ is an AII for (Φ, S, T) then $\{F_\alpha^*\}$ is an AII for (Φ^*, T^*, S^*).

Let $\Phi : \mathcal{H} \to \mathfrak{Y}$ intertwine operators S, S^* with T_1, T_2. Let $\{F_\alpha\} : \mathfrak{Y} \to \mathcal{H}$ be an AII for the intertwining (Φ, S, T_1). It is called a *$*$-approximate inverse intertwining* ($*$-AII) for the ordered pair $((\Phi, S, T_1), (\Phi, S^*, T_2))$ if $\{F_\alpha^* F_\alpha\}$ is an AII for $(\Phi\Phi^*, T_1^*, T_2)$.

A *$*$-approximate inverse semiintertwining* ($*$-AIS) is defined in a similar way: it is an AIS (= AII) $\{F_\alpha\}$ for (Φ, S, T_1) such that $\{F_\alpha^* F_\alpha\}$ is an AIS for $(\Phi\Phi^*, T_1^*, T_2)$.

Theorem 2.5.

(i) *If the pair* $((\Phi, S, T_1),\ (\Phi, S^*, T_2))$ *has a* $*$*-AIS, then*
$$(\operatorname{im} T_1) \cap T_2^{-1}(\Phi\Phi^*(\mathfrak{Y}^*)) \subset \Phi(\mathcal{H}).$$

(ii) *If* $((\Phi, S, T_1),\ (\Phi, S^*, T_2))$ *has a* $*$*-AII, then*
$$||\Phi^{-1}(T_1 y)||^2 = \langle (\Phi\Phi^*)^{-1}(T_2 T_1 y), y\rangle$$
for any $y \in (T_2T_1)^{-1}(\Phi\Phi^*(\mathcal{Y}^*))$.

Corollary 2.6. *If* $((\Phi, S, T_1),\ (\Phi, S^*, T_2))$ *has a* $*$*-AIS then* $\operatorname{im} T_1 \cap \ker T_2 = \{0\}$.

We will finish by a result which has some similarity to Theorem 2.3 but is not related to AII's.

Theorem 2.7. *Let* $\Phi : \mathcal{H} \to \mathfrak{Y}$ *intertwine commuting normal operators* S_1, S_2 *with operators* T_1 *and* T_2 *respectively. Suppose that*
$$\ker(S_1) \cap \Phi^{-1}(T_2\mathfrak{Y}) = \{0\}.$$
Let $\|S_2 h\| \leq \|S_1 h\|$ *for each* $h \in \mathcal{H}$. *Then the inequality* (2.1) *holds for each* $y \in \mathfrak{Y}$ *such that* $T_1 y \in \Phi(\mathcal{H})$ *and* $T_2 y \in \Phi(\mathcal{H})$.

Corollary 2.8. *Let* $\Phi : \mathcal{H} \to \mathfrak{Y}$ *intertwine a normal operator* S *with* T_1 *and its adjoint* S^* *with* T_2. *Suppose that* $\ker(S) \cap \Phi^{-1}(T_2\mathfrak{Y}) = \{0\}$. *Then the inequality* (2.1) *holds for each* $y \in \mathfrak{Y}$ *such that* $T_1 y \in \Phi(\mathcal{H})$ *and* $T_2 y \in \Phi(\mathcal{H})$.

3. General elementary operators: the range

Here by Δ we denote an elementary operator $\sum_{i=1}^n L_{A_i} R_{B_i}$ on $\mathcal{B}(H)$:
$$\Delta(X) = \sum_{i=1}^n A_i X B_i \qquad \text{and set} \qquad \widetilde{\Delta}(X) = \sum_{i=1}^n A_i^* X B_i^*.$$
We denote by Δ_p and $\widetilde{\Delta}_p$ the restriction of Δ and respectively $\widetilde{\Delta}$ to the ideal $\mathcal{S}_p$, $1 \leq p \leq \infty$ (by $\mathcal{S}_\infty$ as usually we denote the ideal $\mathcal{K}(H)$ of all compact operators).

3.1. The intersection with the kernel of adjoint

For each operator acting on a Hilbert space, the closure of its range has zero intersection with the kernel of its adjoint. We will discuss related conditions for elementary operators, that is
$$\overline{\Delta(\mathcal{B}(H))} \cap \ker \tilde{\Delta} = \{0\} \tag{3.1}$$
or a stronger one
$$\|\Delta(X) + Y\| \geq C\|Y\| \text{ for all } X \in \mathcal{B}(H), Y \in \ker \tilde{\Delta} \text{ and some } C > 0, \tag{3.2}$$
(which means that the angle between $\overline{\Delta(\mathcal{B}(H))}$ and $\ker \tilde{\Delta}$ is non-zero), or weaker ones
$$\Delta(\mathcal{B}(H)) \cap \ker \tilde{\Delta} = \{0\} \tag{3.3}$$

and

$$\Delta(\mathcal{S}_p) \cap \ker \tilde{\Delta}_p = \{0\}. \tag{3.4}$$

The last conditions can be rewritten as follows:

$$\ker \tilde{\Delta}\Delta = \ker \Delta \tag{3.5}$$

and

$$\ker \tilde{\Delta}_p\Delta_p = \ker \Delta_p \tag{3.6}$$

respectively. Note that the validity of (3.5) for a class of operators can have strong consequences. For example if it is true for an operator Δ which commutes with $\tilde{\Delta}$ then $\ker \tilde{\Delta} \subset \ker \Delta$. On this way one can obtain various extensions of the Fuglede–Putnam theorem (see Section 4).

Problem 1. Is (3.5) true for inner derivations, that is for operators $\Delta = L_A - R_A$?

The positive answer was known for the case that A is a weighted shift [11] or a subnormal operator with a cyclic vector [11]. Moreover it was shown in [11] that in the latter case the condition (3.1) holds. If A is normal then even (3.2) holds with $C = 1$ [2]. But in general (3.1) is not true for inner derivations: Anderson [3] has shown that the closure of the image of $L_A - R_A$ can contain 1_H.

Note that the intersection of the left part of (3.3) with $\mathcal{S}_1$ is trivial; moreover the following much stronger condition holds:

$$\overline{\Delta(\mathcal{B}(H))}^{w^*} \cap \ker \tilde{\Delta}_1 = \{0\}. \tag{3.7}$$

Here $\overline{\Delta(\mathcal{B}(H))}^{w^*}$ denotes the closure of $\Delta(\mathcal{B}(H))$ in the weak-* topology. To verify (3.7) note that if $X \in \ker \tilde{\Delta}_1$ then $\operatorname{tr}(X^*\Delta(Y)) = \operatorname{tr}(\tilde{\Delta}(X)^*Y) = 0$ for each $Y \in \mathcal{B}(H)$. Hence $\operatorname{tr}(X^*Z) = 0$ for each $Z \in \overline{\Delta(\mathcal{B}(H))}^{w^*}$. It follows that $\operatorname{tr}(X^*X) = 0$ for $X \in \overline{\Delta(\mathcal{B}(H))}^{w^*} \cap \ker \tilde{\Delta}_1$.

It would be important to prove the triviality of the intersection of the left part of (3.3) with $\mathcal{S}_2$:

Problem 2. For which elementary operators Δ does the equality

$$\Delta(\mathcal{B}(H)) \cap \ker \widetilde{\Delta}_2 = 0 \tag{3.8}$$

hold?

We conjecture that for all. But could it be proved at least for inner derivations?

It will be shown in Section 4 that (3.8) is true (in a stronger version) for all normal elementary operators.

Some conditions which are sufficient for (3.5) can be written in terms close in spirit to Voiculescu's notion of quasidiagonality with respect to a symmetrically normed ideal [26]. We restrict ourselves to Schatten ideals $\mathcal{S}_p$ for simplicity.

Let us say that a family $\mathbb{A} = \{A_k : 1 \leq k \leq m\}$ of operators is p-semidiagonal if there exists a sequence of projections P_n of finite rank such that $P_n \to 1$ in the strong operator topology, and

$$\sup_n ||[A_k, P_n]||_p < \infty, \text{ for each } k \leq m.$$

It is clear that if $p_1 < p_2$ then each p_1-semidiagonal family is p_2-semidiagonal. In particular, 1-semidiagonality is the strongest of these conditions. Clearly, any family is ∞-semidiagonal, where $\|\cdot\|_\infty$ is the operator norm.

Let us list some examples of p-semidiagonal families (see [22] for transparent proofs or references).

1) Any family of operators with matrices (with respect to some basis) supported by a finite number of diagonals (i.e., $a_{ij} = 0$ if $|i-j| > m$ for some $m \in \mathbb{N}$) is 1-semidiagonal. One of the simplest classes of such examples consists of families of weighted shifts.

This class of examples can be considerably extended as follows.

If (a_{ij}) is a matrix of an operator A in a basis $\{e_n\}_{n=1}^\infty$, let us set $|A|_n = \sup_{|i-j|=n} |a_{ij}|$ and $|A|_{\text{diag}} = \sum_{n=1}^\infty n|A|_n$. We say that A is diagonally bounded (with respect to the basis $\{e_n\}_{n=1}^\infty$) if $|A|_{\text{diag}} < \infty$. Then:

2) Any family of diagonally bounded (with respect to the same basis) operators is 1-semidiagonal.

As a consequence we obtain the following result of Voiculescu [26]:

3) Any family which belongs to the algebra of operators on $L_2(\mathbb{T})$ generated by shifts $u(t) \mapsto u(t-\theta)$ and multiplication operators by twice differentiable functions $f \in C^2(\mathbb{T})$ (we will call the elements of these algebras "generalized Bishop's operatos"), is 1-semidiagonal.

To deduce this from 2), it suffices to calculate the matrices of shift and multiplication operators in the standard basis $e_k = \exp(ikt)$, $k \in \mathbb{Z}$.

4) It is important that all normal operators of finite multicyclicity are 2-semidiagonal. More generally, if $f_1, \ldots, f_n$ are Lipschitz functions on the spectrum of a normal operator A, then the family $(f_1(A), \ldots, f_n(A))$ is p-semidiagonal where p is the Hausdorff dimension of $\sigma(A)$.

Slightly more general, if the essential dimension of a family $\mathbb{A} = (A_1, \ldots, A_n)$ of commuting normal operators does not exceed $p \leq 2$ then $\mathbb{A}$ is p-semidiagonal.

5) Any multicyclic almost normal operator A is 2-semidiagonal ([25]).

Recall that an operator A is *almost normal* if its self-commutator $[A^*, A]$ is nuclear.

Theorem 3.1. *If the left coefficient family $\mathbb{A}$ of an elementary operator Δ is 1-semidiagonal then the equality* (3.3) *holds.*

If $\mathbb{A}$ is $p/(p-2)$-semidiagonal then (3.4) *holds.*

Of course one could impose the same restriction (of 1-semidiagonality) on the right coefficient family $\mathbb{B}$ of Δ.

Problem 3. Find a condition which involves both coefficient families so that (3.3) or (3.4) is valid. In particular, does (3.3) hold if both $\mathbb{A}$ and $\mathbb{B}$ are 2-semidiagonal?

To outline the proof of Theorem 3.1, note that the operators Δ and Δ_p are intertwined by the injection $\Phi_p : \mathcal{S}_p \to \mathcal{B}(H)$. Similarly the injection $\Phi_{p_1,p_2} : \mathcal{S}_{p_1} \to \mathcal{S}_{p_2}$, intertwines Δ_{p_1} and Δ_{p_2} for $p_1 \leq p_2$. If $\mathbb{A}$ is p-semidiagonal then one can use the projections P_n from the definition of semidiagonality, to construct an AII for these intertwinings by setting $F_n(X) = P_n X$. We describe firstly the results that relate properties of Δ and Δ_2: this is important because the latter acts on the Hilbert space $\mathcal{H} = \mathcal{S}_2$ and its adjoint is $\widetilde{\Delta}_2$.

Theorem 3.2.

(i) *If the left coefficient family $\mathbb{A}$ of Δ is 2-semidiagonal then there exists an AII for $(\Phi_2, \Delta_2, \Delta)$.*

(ii) *If $\mathbb{A}$ is 1-semidiagonal then*

  (a) *there exists a *-AIS for $(\Phi_2, \Delta_2, \Delta)$ and $(\Phi_2, \widetilde{\Delta}_2, \widetilde{\Delta})$;*

  (b) *there exists an AII for $(\Phi_2, \Delta_2, \Delta_\infty)$;*

  (c) *there exists a *-AII for $(\Phi_2, \Delta_2, \Delta_\infty)$ and $(\Phi_2, \widetilde{\Delta}_2, \widetilde{\Delta}_\infty)$.*

(iii) *If $\mathbb{A}$ is $p/(p-2)$-semidiagonal, $p > 2$, then there exists a *-AII for $(\Phi_{2,p}, \Delta_2, \Delta_p)$ and $(\Phi_{2,p}, \widetilde{\Delta}_2, \widetilde{\Delta}_p)$*

Now to prove Theorem 3.1 it suffices to apply Theorem 3.2 (parts (ii-a) and (iii)) and Corollary 2.6.

3.2. Hyponormality

Let us say that an elementary operator Δ is *formally positive* if $\operatorname{tr}(\Delta(X)X^*) \geq 0$ for each $X \in \mathcal{S}_2$. Furthermore Δ is *formally hyponormal* if $\widetilde{\Delta}\Delta - \Delta\widetilde{\Delta}$ is formally positive.

Theorem 3.3. *Let Δ be formally hyponormal and let its left coefficient family $\mathbb{A}$ be 2-semidiagonal.*

(i) *If $\Delta(X) \in \mathcal{S}_2$, for some $X \in \mathcal{B}(H)$, then $\widetilde{\Delta}(X) \in \mathcal{S}_2$ and*

$$\|\Delta(X)\|_2 \geq \|\widetilde{\Delta}(X)\|_2.$$

As a consequence we get

(ii) $\ker \Delta \subset \ker \widetilde{\Delta}$.

Indeed, by Theorem 3.2 (i), the assumption of 2-semidiagonality implies the existence of AII for $(\Phi_2, \Delta_2, \Delta)$. Since formal hyponormality of Δ means that Δ_2 is a hyponormal operator, the inequality $\|\widetilde{\Delta}(X)\|_2 \leq \|\Delta(X)\|_2$ holds for each operator $X \in \mathcal{S}_2$. It remains to apply Theorem 2.3 to the intertwinings $(\Phi_2, \Delta_2, \Delta)$ and $(\Phi_2, \widetilde{\Delta}_2, \widetilde{\Delta})$.

Considering elementary operators of the form $\Delta = L_A + R_B$ it is easy to see that $\widetilde{\Delta}\Delta - \Delta\widetilde{\Delta} = L_{[A^*,A]} + R_{[B,B^*]}$. Therefore such operator is formally hyponormal if A and B^* are hyponormal. One can also show that the converse is also true. Note that if a hyponormal operator A is multicyclic then its selfcommutator $[A^*, A]$ is nuclear ([4]) so A is almost normal. Taking into account that multicyclic normal operators are 2-semidiagonal we apply Theorem 3.3 and obtain

Corollary 3.4. *Let operators A and B^* be hyponormal and multicyclic. Then*

$$\|AX - XB\|_2 \geq \|A^*X - XB^*\|_2 \tag{3.9}$$

for each operator X such that $AX - XB \in \mathcal{S}_2$.

This result from [22] was proved earlier under more restrictive conditions on A, B and X (see [7] and numerous references therein).

Theorem 3.3 is related to the following result which is not restricted by hyponormality assumptions.

Theorem 3.5. *If $\mathbb{A}$ is 1-semidiagonal then*

$$\|\Delta(X)\|_2^2 = \mathrm{tr}(X^*\widetilde{\Delta}\Delta(X)) \tag{3.10}$$

for each compact operator X such that $\widetilde{\Delta}\Delta(X) \in \mathcal{S}_1$.

This statement can be proved in the same way as the previous results by using Theorem 2.5 (ii) and Theorem 3.2 (ii-c).

To see its relation to Theorem 3.3, note that if Δ is formally hyponormal then clearly $\mathrm{tr}(X^*\widetilde{\Delta}\Delta(X)) \geq \mathrm{tr}(X^*\Delta\widetilde{\Delta}(X))$.

3.3. Ranges of derivations and traces of commutators

Now we will study which trace class operators can belong to the range of an elementary operator Δ. For this we need information about AII's for the intertwinings $(\Phi_1, \Delta_1, \Delta)$ and $(\Phi_{1,p}, \Delta_1, \Delta_p)$.

It will be convenient, for each $p \in (1;\infty)$, to denote by p' the number $\frac{p}{p-1}$, and set $p' = 1$ if $p = \infty$, $p' = \infty$ if $p = 1$.

Theorem 3.6.

(i) *If the left coefficient family $\mathbb{A}$ of Δ is 1-semidiagonal then there exists an AIS for $(\Phi_1, \Delta_1, \Delta)$.*

(ii) *If $\mathbb{A}$ is p'-semidiagonal then there exists an AII for $(\Phi_{1,p}, \Delta_1, \Delta_p)$.*

Applying Theorem 2.1 we deduce from part (ii) of Theorem 3.6

Corollary 3.7. *If $\mathbb{A}$ is p'-semidiagonal then $\mathcal{S}_1 \cap \Delta(\mathcal{S}_p)$ is contained in the $\|\cdot\|_1$-closure of $\Delta(\mathcal{S}_1)$.*

In particular, we have the following consequence.

Corollary 3.8. *Let the coefficients of Δ satisfy the condition $\sum_{i=1}^n B_iA_i = 0$. If $\mathbb{A}$ is p'-semidiagonal then $\mathrm{tr}(\Delta(X)) = 0$ for each operator $X \in \mathcal{S}_p$ for which $\Delta(X) \in \mathcal{S}_1$.*

Indeed if $Y \in \mathcal{S}_1$ then $\mathrm{tr}(\Delta(Y)) = \mathrm{tr}(\sum_{i=1}^n B_iA_iY) = 0$, so the result follows from Corollary 3.7 and the continuity of trace on $\mathcal{S}_1$.

The result can be applied to the problem "when the trace of a commutator is equal to 0?". It is well known that a commutator $[A, X]$ has zero trace if X is nuclear or if both operators A and X are Hilbert–Schmidt. Weiss [27] proved that the same is true if X is Hilbert–Schmidt and A is normal. The following proposition, which is an easy consequence of Corollary 3.8, widely extends this result.

Corollary 3.9. *Let $p > 1$. Suppose that $\{A_k\}_{k=1}^n$ is p'-semidiagonal, $X_k \in \mathcal{S}_p$ and $\sum_{k=1}^n [A_k, X_k] \in \mathcal{S}_1$. Then*

$$\operatorname{tr}\left(\sum_{k=1}^n [A_k, X_k]\right) = 0.$$

For $p = \infty$, this gives, for instance, that the commutator of a Hermitian operator (or a weighted shift, or a Bishop's operator) with a compact operator has zero trace if nuclear. Taking $p = 2$, we obtain the same for the commutator of a normal (or almost normal) multicyclic operator with a Hilbert–Schmidt operator. The restriction on multicyclicity can be easily removed.

Choosing in Proposition 3.9 for A_k the multiplication operators M_{f_k} on $L_2([0,1])$ and for X_k the integral operators with kernels $F_k \in L_2([0,1]^2)$, one obtains the following result:

Corollary 3.10. *If $f_k \in Lip_{1/2}([0,1])$, $1 \leq k \leq n$, then there are no functions $F_k \in L_2([0,1]^2)$ satisfying the condition*

$$\sum_{k=1}^n (f_k(x) - f_k(y)) F_k(x, y) = 1. \tag{3.11}$$

It was asked by Weiss if this is true for $f_k \in C[0,1]$; the answer is negative (see [22, Proposition 8.7] which shows that $1/2$ in Corollary 3.10 cannot be even changed by $1/3$).

The constant p' in Proposition 3.9 is sharp for $p = 2$, i.e., the condition $X_k \in \mathcal{S}_2$ cannot be weakened to $X_k \in \mathcal{S}_q$ for any $q > 2$ (see [22, Example 8.5]).

Problem 4. Is the constant p' sharp for all $p \geq 1$?

4. Normal elementary operators

In this section the coefficient families $\mathbb{A}, \mathbb{B}$ of an elementary operator Δ are assumed to be commutative and to consist of normal operators. Such elementary operators are called normal.

4.1. Spectral subspaces

It is well known (see for example a much more general result of [6]) that

$$\sigma(\Delta) = \left\{\sum_{k=1}^n \lambda_i \mu_i : \lambda \in \sigma(\mathbb{A}),\ \mu \in \sigma(\mathbb{B})\right\}. \tag{4.1}$$

We will study here some other spectral characteristics of elementary operators.

Recall that an operator T on a Banach space $\mathcal{X}$ is decomposable if to each compact set $\alpha \subset \sigma(T)$ there corresponds a closed subspace $E_T(\alpha)$ invariant for all operators commuting with T and satisfying the following conditions:

(a) $\sigma(T|E_T(\alpha)) \subset \alpha$;

(b) If U, V are open subsets of $\mathbb{C}$ and $U \cup V \supset \sigma(T)$ then there are compact sets $\alpha \subset U$, $\beta \subset V$ with $E_T(\alpha) + E_T(\beta) = \mathcal{X}$.

The subspaces $E_T(\alpha)$ are called *spectral subspaces* of T; the map $\alpha \to E_T(\alpha)$ is *the spectral capacity* of T.

To describe spectral subspaces $E_\Delta(\alpha)$ of a normal elementary operator Δ we need the following notion. Let $E_\mathbb{A}(\cdot)$ and $E_\mathbb{B}(\cdot)$ be the spectral measures of the coefficient families $\mathbb{A}$ and $\mathbb{B}$ of Δ on $\sigma(\mathbb{A})$ and $\sigma(\mathbb{B})$ respectively. An operator $X \in \mathcal{B}(H)$ is said to be supported by a set $M \subset \sigma(\mathbb{A}) \times \sigma(\mathbb{B})$ if $E_\mathbb{A}(U) X E_\mathbb{B}(V) = 0$ for every Borel rectangle $U \times V \subset \sigma(\mathbb{A}) \times \sigma(\mathbb{B})$ non-intersecting M. We will write in this case $\operatorname{supp} X \subset M$.

It is not difficult to check that for each subset $M \subset \sigma(\mathbb{A}) \times \sigma(\mathbb{B})$, the subspace H_M of all operators supported by M is invariant for Δ. Moreover it can be deduced in the same way as (4.1) that $\sigma(\Delta|H_M) \subset \{\sum_{k=1}^n \lambda_i \mu_i : (\lambda, \mu) \in M\}$.

For every compact $\alpha \subset \sigma(\Delta)$, we set

$$M_{\mathbb{A},\mathbb{B}}(\alpha) = \Big\{(\lambda, \mu) \in \sigma(\mathbb{A}) \times \sigma(\mathbb{B}) : \sum_{k=1}^n \lambda_k \mu_k \in \alpha\Big\}.$$

Set now

$$E_\Delta(\alpha) = \{X \in \mathcal{B}(H) : \operatorname{supp} X \subset M_{\mathbb{A},\mathbb{B}}(\alpha)\}.$$

In other words $E_\Delta(\alpha) = H_{M_{\mathbb{A},\mathbb{B}}(\alpha)}$.

Theorem 4.1. [18] *A normal elementary operator Δ is decomposable; the map $\alpha \to E_\Delta(\alpha)$ is its spectral capacity.*

In particular the space $E_\Delta(\{0\})$ of all operators supported by $M_{\mathbb{A},\mathbb{B}}(0)$ coincides with $\{X : \|\Delta^n(X)\|^{1/n} \to 0\}$; in other words it is *the root space* of Δ.

It is important that the set $M_{\mathbb{A},\mathbb{B}}(\{0\})$ has a comparatively simple structure: it is the union of a null set and a countable family of rectangles. More precisely, let $\mu_\mathbb{A}$, $\mu_\mathbb{B}$ be scalar spectral measures of $\mathbb{A}$ and $\mathbb{B}$, and let $m = \mu_\mathbb{A} \times \mu_\mathbb{B}$ be the product measure on $\sigma(\mathbb{A}) \times \sigma(\mathbb{B})$.

Lemma 4.2. *There are measurable sets $A_i \subset \sigma(\mathbb{A})$, $B_i \subset \sigma(\mathbb{B})$, $1 \le i < \infty$, and an m-null subset C of $\sigma(\mathbb{A}) \times \sigma(\mathbb{B})$ such that $M_{\mathbb{A},\mathbb{B}}(0) = (\cup_{i=1}^\infty A_i \times B_i) \cup C$.*

It follows from Lemma 4.2 that the space of all Hilbert–Schmidt operators in $E_\Delta(\{0\})$ is generated by the union of the spaces of all Hilbert–Schmidt operators supported by rectangles $A_i \times B_i$. Using this one obtains easily the following result:

Corollary 4.3. *The space $\mathcal{S}_1 \cap E_\Delta(\{0\})$ is dense in $\mathcal{S}_2 \cap E_\Delta(\{0\})$ with respect to the norm of the ideal $\mathcal{S}_2$.*

Since for a normal operator on a Hilbert space the root space coincides with the kernel, the result can be reformulated as follows: $\ker \Delta_1$ is a dense subspace of $\ker \Delta_2$.

Now we can show that for normal elementary operators the equality (3.8) holds in a much stronger version.

Let $\overline{M}^{w^*}$ denote the weak*-closure of a set $M \subset \mathcal{B}(H)$.

Theorem 4.4. *For every normal elementary operator* Δ,

$$\overline{\Delta(\mathcal{B}(H))}^{w^*} \cap \ker \widetilde{\Delta}_2 = 0. \tag{4.2}$$

Indeed it is trivial to check that $\operatorname{tr}(\Delta(X)Y) = 0$ for each $X \in \mathcal{B}(H)$ and $Y \in \ker \widetilde{\Delta}_1$. Hence by continuity $\operatorname{tr}(ZY) = 0$ for $Z \in \overline{\Delta(\mathcal{B}(H))}^{w^*}$.

If also $Z \in \mathcal{S}_2$ then, again by continuity, $\operatorname{tr}(ZY) = 0$ for all Y from the closure of $\ker \widetilde{\Delta}_1$ in $\mathcal{S}_2$. But by Corollary 4.3, this closure coincides with $\ker \widetilde{\Delta}_2$, which in its turn equals $\ker \Delta_2$ because Δ_2 is a normal operator and $\widetilde{\Delta}_2$ is its adjoint. So if $Z \in \overline{\Delta(\mathcal{B}(H))}^{w^*} \cap \ker \widetilde{\Delta}_2$ then $Z = 0$.

Note that we proved a more general statement:

$$\overline{\Delta(\mathcal{B}(H))}^{w^*} \cap \mathcal{S}_2 \text{ is orthogonal to } \ker \widetilde{\Delta}_2. \tag{4.3}$$

From this one can easily deduce the following result of Turnšek [24]:

Proposition 4.5. *For every normal elementary operator* Δ *the intersection of its range with* $\mathcal{S}_2$ *is contained in the* $\|\cdot\|_2$*-closure of* $\Delta(\mathcal{S}_2)$.

Indeed the $\|\cdot\|_2$-closure of $\Delta(\mathcal{S}_2)$ is just the orthogonal complement of $\ker \widetilde{\Delta}_2$.

Problem 5. For which $p > 2$ does the equality (4.2) hold with $\tilde{\Delta}_p$ instead of $\tilde{\Delta}_2$?

Remark 4.6. We note that $\overline{\Delta(\mathcal{B}(H))}^{w^*} \cap \ker \widetilde{\Delta} = 0$ does not hold in general even for normal inner derivations. One can easily construct $\Delta = L_A - R_A$ such that $\overline{\Delta(\mathcal{B}(H))}^{w^*} = \mathcal{B}(H)$.

4.2. Ascent

The relation between the kernel of an operator T and its root space is a subject of the so-called *theory of thin spectral structure.* Since the root space contains the kernels of all operators T^n, a natural intermediate question is the interrelations of these spaces. If the chain $\ker T \subset \ker T^2 \subset \cdots$ stabilizes on number m: $\ker T^m = \ker T^{m+1} = \cdots$ then we say that the ascent of T equals m.

We will see that the ascent m of a normal elementary operator can be estimated via the essential dimension of its left coefficient family, and that $E_\Delta(0) = \ker \Delta^m$.

Theorem 4.7. *If* ess-dim$\mathbb{A} \le r$ *then* $E_\Delta(0) = \ker \Delta^{(r/2]}$ *where* $(r/2] = [(r+1)/2]$, *the minimal integer* $\ge r/2$

For operators of the form $X \to AX + XB$ this result was obtained by Anderson and Foias [2]; clearly in this case it states that $E_\Delta(0) = \ker \Delta$.

The first step of the proof is to show that $\|\Delta|E_\Delta(0)\| \leq 2|\sigma(\mathbb{B})| \operatorname{diam}\sigma(\mathbb{A})$, where by $|\sigma(\mathbb{B})|$ we mean $\sup\{|\lambda| : \lambda \in \sigma(\mathbb{B})\}$. Hence

$$\|\Delta^k|E_\Delta(0)\| \leq (2|\sigma(\mathbb{B})| \operatorname{diam}\sigma(\mathbb{A}))^k$$

for each k.

If γ is a compact subset of $\sigma(\mathbb{A})$ then changing Δ by $\Delta R_{E_{\mathbb{A}}(\gamma)}$ we obtain the inequality $\|\Delta^k R_{E_{\mathbb{A}}(\gamma)}|E_\Delta(0)\| \leq (2|\sigma(\mathbb{B})| \operatorname{diam}\gamma)^k$. Hence taking $X \in E_\Delta(0)$ and setting $T = \Delta^k(X)$ we obtain:

$$\|TE_{\mathbb{A}}(\gamma)\| \leq C(\operatorname{diam}\gamma)^k \tag{4.4}$$

for all $\gamma \subset \sigma(\mathbb{A})$.

For each finite covering of $\sigma(\mathbb{A})$ by disjoint Borel sets $\gamma_1, \ldots, \gamma_N$, one can consider an orthogonal system $\{e_1, \ldots, e_N\}$ of unit vectors e_j supported in $E_{\mathbb{A}}(\gamma_j)$; the condition (4.4) gives that $\|TP_N\|_2^2 \leq C\sum_{j=1}^N (\operatorname{diam}\gamma_j)^{2k}$ where P_N is the projection onto the subspace generated by $\{e_1, \ldots, e_N\}$. If $\mathbb{A}$ is cyclic (this can be assumed without reducing generality) then such orthogonal systems can approximate a basis. It follows that if the Hausdorff dimension of $\sigma(\mathbb{A})$ does not exceed $2k$ then each operator $T \in \mathcal{B}(H)$ satisfying the condition (4.4) belongs to $\mathcal{S}_2$.

We get that $\Delta^k(X) \in \mathcal{S}_2$ for all $X \in E_\Delta(0)$. Since $E_\Delta(0)$ is invariant for Δ, $\Delta^k(X) \in \mathcal{S}_2 \cap E_\Delta(0) = \ker\Delta_2 = \ker\widetilde{\Delta}_2$. By Theorem 4.4, $\Delta^k(X) = 0$.

4.3. Fuglede type theorems

The famous Fuglede–Putnam theorem says that $\ker\Delta = \ker\tilde{\Delta}$ if Δ is an operator of the form $X \mapsto N_1X - XN_2$, where N_1, N_2 are normal operators. A natural question arising in this context is whether for every normal elementary operator Δ the equations

$$\Delta(X) = 0 \tag{4.5}$$

and

$$\widetilde{\Delta}(X) = 0 \tag{4.6}$$

are equivalent. This question was answered negatively in [19]. The construction of the counterexample is based on a modification of the famous L. Schwartz example of a set of spectral non-synthesis for the Fourier algebra of the group $\mathbb{R}^3$.

It is clear that $E_\Delta(0) = H_{M_{\mathbb{A},\mathbb{B}}} = E_{\widetilde{\Delta}}(0)$. Applying Theorem 4.7 we obtain the following version of the Fuglede-Putnam theorem:

Theorem 4.8. *If* ess-dim$\mathbb{A} \leq r$ *then* $\ker\Delta^{(r/2]} = \ker\widetilde{\Delta}^{(r/2]}$.

The following consequence of Theorem 4.8 can be also immediately deduced from Theorem 3.3 (ii).

Theorem 4.9. *If* ess-dim$\mathbb{A} \leq 2$ *then equations* (4.5) *and* (4.6) *are equivalent.*

It is reasonable to regard the work around Fuglede Theorem in a wider way: as the study of conditions for the coincidence or inclusion of kernels of elementary operators. In other words: which linear operator equations are equivalent, how the solution spaces of different equations are related? The following theorem gives information for the case of 2-dimensional coefficient families.

Theorem 4.10. *Let* ess-dim$\mathbb{A} \leq 2$. *Let* $\{f_i\}_{i=1}^m$ *be Lipschitz functions on* $\sigma(\mathbb{A})$, $\{g_i\}_{i=1}^m$ *be Borel functions on* $\sigma(\mathbb{B})$. *Assume that* $\sum_{i=1}^m f_i(\lambda)g_i(\mu) = 0$ *for each* $(\lambda, \mu) \in \sigma(\mathbb{A}) \times \sigma(\mathbb{B})$ *satisfying the condition* $\sum_{k=1}^n \lambda_k \mu_k = 0$. *Then each solution* X *of the linear operator equation*

$$\sum_{k=1}^n A_k X B_k = 0 \tag{4.7}$$

satisfies the equation

$$\sum_{i=1}^m f_i(\mathbb{A}) X g_i(\mathbb{B}) = 0. \tag{4.8}$$

Clearly Theorem 4.9 is a special case of Theorem 4.10 – it corresponds to the functions $f_i(\lambda) = g_i(\lambda) = \overline{\lambda_i}$.

To deduce Theorem 4.10 from Theorem 4.7, denote by $f(\mathbb{A})$ and $g(\mathbb{B})$ respectively the families $\{f_i(\mathbb{A}) : 1 \leq i \leq m\}$ and $\{g_i(\mathbb{B}) : 1 \leq i \leq m\}$. Let $\Delta' = \sum_{i=1}^m L_{f_i(\mathbb{A})} R_{g_i(\mathbb{B})}$. In our assumptions $M_{\mathbb{A},\mathbb{B}}(0) \subset M_{f(\mathbb{A}),g(\mathbb{B})}(0)$ whence $E_\Delta(0) \subset E_{\Delta'}(0)$. Furthermore ess-dim $f(\mathbb{A}) \leq 2$, so $E_{\Delta'}(0) = \ker \Delta'$ by Theorem 4.7.

Clearly the "Fuglede theorem", for arbitrary normal Δ, holds in $\mathcal{S}_2$ (that is the equations (4.5) and (4.6) are equivalent in $\mathcal{S}_2$) and hence in $\mathcal{S}_p$, $p < 2$.

Problem 6. Let Δ be a normal elementary operator. Are equations (4.5) and (4.6) equivalent in $\mathcal{S}_p$, $2 < p \leq \infty$?

One can ask about the validity of similar results in C^*-algebras different from $\mathcal{B}(H)$. Namely, let $\mathcal{A}$ be a C^*-algebra. Consider elementary operators Δ on $\mathcal{A}$, i.e., $\Delta(c) = \sum_{k=1}^n a_k c b_k$, where $a_k, b_k \in \mathcal{A}$. We say that Δ is normal if $\{a_k\}_{k=1}^n$ and $\{b_k\}_{k=1}^n$ are families of commuting normal elements in $\mathcal{A}$.

Let us call a C*-algebra $\mathcal{A}$ 1-Fuglede (or (F1), for short) if $\ker \tilde{\Delta} = \ker \Delta$ for any normal elementary operator Δ. If $\ker \Delta^2 = \ker \Delta$ for any normal Δ then $\mathcal{A}$ is said to be 2-Fuglede (F2). If $\ker \tilde{\Delta}\Delta = \ker \Delta$ for any Δ then $\mathcal{A}$ is said to be 3-Fuglede (F3). Clearly (F3) $\Longrightarrow$ (F2) $\Longrightarrow$ (F1).

It follows from the above discussion that the algebra $\mathcal{B}(H)$ is not (F1) (for any infinite-dimensional H). It is not difficult to see that

(i) the algebra, $\mathcal{S}_\infty$, of compact operators is (F2);
(ii) as a consequence each CCR-algebra is (F2);
(iii) each C*-algebra with a faithful family of traces (for example a simple unital hyperfinite C*-algebra) is 3-Fuglede.

Problem 7. Which C^*-algebras $\mathcal{A}$ are F1, F2 and F3? Is $\mathcal{S}_\infty$ 3-Fuglede? Is any GCR-algebra (F2)? What about the Calkin algebra $\mathcal{B}(H)/\mathcal{S}_\infty$?

We will briefly list other extensions of the Fuglede Theorem.

E.A. Gorin [8] discovered that if in the classical Rosenblum's proof of the Fuglede Theorem one uses instead of the Liouville theorem the Phragmen–Lindelöf Theorem, then the result extends to a more general class of operators than the normal ones.

Theorem 4.11. *Let an operator A, acting on a Banach space $\mathfrak{X}$, be decomposed into a sum $A = T + iS$, where $[S,T] = 0$, S has real spectrum and T satisfies the condition*

$$\|\exp(itT)\| = o(|t|) \text{ for t} \to \infty. \tag{4.9}$$

Then $\ker A = \ker T \cap \ker S$.

Applying this theorem to multiplication operators we obtain

Corollary 4.12. *Let an operator N on a Banach space $\mathfrak{X}$ can be written as a sum $N = U + iV$ where $[U,V] = 0$, V has real spectrum and U satisfies condition*

$$\|\exp(itU)\| = o(|t|^{1/2}). \tag{4.10}$$

Set $N' = U - iV$. Then the equations $NX - XN = 0$ and $N'X - XN' = 0$ are equivalent.

An operator T on a Banach space is called *Hermitian* if $\|\exp(itT)\| = 1$ for all $t \in \mathbb{R}$.

The following result belongs to K.Boyadzhiev [5].

Theorem 4.13. *Let $P(t_1, \dots, t_n)$ be a polynomial in n variables without zeros in $\mathbb{R}^n \backslash \{0\}$. If $T_1, \dots, T_n$ are commuting Hermitian operators then* $\ker P(T_1, \dots, T_n) \subset \cap_{k=1}^n \ker T_k$.

To deduce the Fuglede Theorem from Theorem 4.13, one should take $P(t_1, t_2) = t_1 + it_2$, $T_1 = L_A - R_A$, $T_2 = L_B - R_B$ where A, B are the real and imaginary parts of a normal operator $N \in \mathcal{B}(H)$.

E.A. Gorin [8] obtained also the following "non-commutative" version of the Fuglede Theorem.

Let $\mathcal{F}_2$ be the free algebra with generators a, b. Consider the formal power series $f(z) = \exp(za)\exp(zb)$ with coefficients in $\mathcal{F}_2$. Then there is a formal power series $g(z)$ with coefficients in $\mathcal{F}_2$ such that $f(z) = \exp(g(z))$. Let us denote by $c_k(a,b)$ the coefficients of this power series:

$$g(z) = c_0(a,b) + c_1(a,b)z + \cdots .$$

It is easy to obtain explicit formulas for $c_k(a,b)$ for small k. For example $c_0(a,b) = 1$, $c_1(a,b) = a + b$, $c_2(a,b) = [a,b]/2$.

Since $c_k(a,b)$ are elements of $\mathcal{F}_2$ ("non-commutative polynomials"), one can calculate $c_k(A,B)$ for each pair A, B of elements of any algebra.

Theorem 4.14. *Let T, S be operators on a Banach space $\mathfrak{X}$ such that T satisfies (4.10), S has real spectrum, and let $X \in B(\mathfrak{X})$. If $[c_k(T, iS), X] = 0$ for all $k \in \mathbb{N}$ then $[T,X] = [S,X] = 0$.*

To see that this theorem extends Corollary 4.12 note that if $[a,b]=0$ then $c_k(a,b)=0$ for all $k\geq 2$. Hence in this case the assumption becomes just $[c_1(a,ib),x]=0$, that is $a+ib$ commutes with x.

4.4. Norm inequalities

The following result is a special case of Theorem 3.3 (i).

Theorem 4.15. *Suppose that* ess-dim$\mathbb{A}\leq 2$. *If* $\Delta(X)\in\mathcal{S}_2$ *then* $\widetilde{\Delta}(X)\in\mathcal{S}_2$ *and*

$$\|\Delta(X)\|_2=\|\widetilde{\Delta}(X)\|_2. \tag{4.11}$$

It extends the famous result of Gary Weiss [29] which states that (4.11) holds for $\Delta=L_{N_1}-R_{N_2}$ where N_i are normal operators.

The arguments similar to ones used in Theorem 3.3 allow to extend Theorem 4.15 in spirit of Theorem 4.10:

Theorem 4.16. *Let* ess-dim$\mathbb{A}\leq 2$. *Let* $\{f_i\}_{i=1}^m$ *be Lipschitz functions on* $\sigma(\mathbb{A})$, $\{g_i\}_{i=1}^m$ *be Borel functions on* $\sigma(\mathbb{B})$. *Assume that* $|\sum_{i=1}^m f_i(\lambda)g_i(\mu)|\leq|\sum_{k=1}^n\lambda_k\mu_k|$ *for each* $(\lambda,\mu)\in\sigma(\mathbb{A})\times\sigma(\mathbb{B})$.

If $\sum_{k=1}^n A_kXB_k\in\mathcal{S}_2$, *for some* $X\in\mathcal{B}(H)$, *then* $\sum_{i=1}^m f_i(\mathbb{A})Xg_i(\mathbb{B})\in\mathcal{S}_2$ *and*

$$\left\|\sum_{i=1}^m f_i(\mathbb{A})Xg_i(\mathbb{B})\right\|_2\leq\left\|\sum_{k=1}^n A_kXB_k\right\|_2.$$

Kittaneh [14] established the following special case of this result: if A is a normal operator and f is a Lipschitz function on $\sigma(A)$ then

$$\|f(A)X-Xf(A)\|_2\leq k(f)\|AX-XA\|_2 \tag{4.12}$$

where $k(f)$ is the Lipschitz constant of f.

Without restrictions on the dimension of spectra the equality (4.11) can fail because even the kernels of Δ and $\widetilde{\Delta}$ can differ. Nevertheless Weiss [28] proved the following remarkable result:

Theorem 4.17. *Let* Δ *be a normal elementary operator. If both* $\Delta(X)$ *and* $\widetilde{\Delta}(X)$ *are Hilbert–Schmidt operators then* (4.11) *holds.*

One can prove Theorem 4.17 applying Corollary 2.8. Take as usually $\mathcal{H}=\mathcal{S}_2$, $S=\Delta_2$, $\mathfrak{Y}=\mathcal{B}(H)$, $T_1=\Delta$, $T_2=\widetilde{\Delta}$, then the identity inclusion $\Phi_2:\mathcal{S}_2\to\mathcal{B}(H)$ intertwines the operators. The condition $\ker S\cap\Phi^{-1}(T_2\mathfrak{Y})=0$, which is just $\ker\Delta_2\cap\widetilde{\Delta}(\mathcal{B}(H))=0$, holds by Theorem 4.4. So inequality (2.1) holds which means that $\|\widetilde{\Delta}(X)\|_2\leq\|\Delta(X)\|_2$ if $\Delta(X)\in\mathcal{S}_2$ and $\widetilde{\Delta}(X)\in\mathcal{S}_2$. Interchanging $\widetilde{\Delta}$ and Δ we obtain (4.11).

Using Theorem 2.7 instead of Corollary 2.8 one can obtain the following more general result:

Theorem 4.18. *Let* $\{f_i\}_{i=1}^m$ *be Lipschitz functions on* $\sigma(\mathbb{A})$, $\{g_i\}_{i=1}^m$ *be Borel functions on* $\sigma(\mathbb{B})$. *Assume that* $|\sum_{i=1}^m f_i(\lambda)g_i(\mu)|\leq|\sum_{k=1}^n\lambda_k\mu_k|$ *for each* $(\lambda,\mu)\in\sigma(\mathbb{A})\times\sigma(\mathbb{B})$.

If $\sum_{k=1}^{n} A_k X B_k \in \mathcal{S}_2$ *and* $\sum_{i=1}^{m} f_i(\mathbb{A}) X g_i(\mathbb{B}) \in \mathcal{S}_2$, *for some* $X \in \mathcal{B}(H)$, *then*

$$\left\| \sum_{i=1}^{m} f_i(\mathbb{A}) X g_i(\mathbb{B}) \right\|_2 \leq \left\| \sum_{k=1}^{n} A_k X B_k \right\|_2 .$$

The mentioned result of (4.11) extends to other Schatten ideals: for every $p \in (1, \infty)$ there is a constant $c = c_p > 0$ such that

$$\|A^*X - XA^*\|_p \leq c\|AX - XA\|_p \tag{4.13}$$

for all normal operators A and all operators X (Abdessemed and Davies [1], for $p > 2$, Shulman [20], for $p < 2$). This is not the case for $p = 1$ and $p = \infty$ (Yu.B. Farforovskaya [10]). Kissin and Shulman [12] proved that the statement remains true for the operator norm and $\mathcal{S}_1$-norm, if one imposes a restriction on spectra of normal operators. Namely, for each C^2-smooth Jordan line L there is a constant $c = c_L$ such that

$$\|A^*X - XA^*\| \leq c\|AX - XA\| \tag{4.14}$$

if $\sigma(A) \subset L$. It was also proved in [12] that a kind of smoothness of spectrum is necessary for the validity of (4.14): if for some normal operator A the inequality (4.14) holds for all X, then given a sequence $\lambda_n \in \sigma(A)$ converging to $\lambda \in \sigma(A)$, there is a limit $\lim(\lambda_n - \lambda)/|\lambda_n - \lambda|$.

Potapov and Sukochev [15] using a powerful technique of Banach space geometry and harmonic analysis proved that (4.12) extends to all S_p ideals, if $A = A^*$. For all normal A this was proved in [13].

If we note that for each X, the map $\delta_X : A \mapsto AX - XA$ is a derivation of the algebra $\mathcal{B}(H)$ then the question arises if the above results can be extended to derivations of more general class (defined on subalgebras of $\mathcal{B}(H)$). The following theorem was proved in [20].

Theorem 4.19. *Let* δ *be a derivation from a *-subalgebra* $D(\delta) \subset \mathcal{B}(H)$ *to* $\mathcal{B}(H)$, *and let* $A \in D(\delta)$ *be a normal operator without eigenvalues. Then*

(i) *If* $\delta(A) \in \mathcal{S}_p$ *and* $\delta(A^*) \in K(H)$ *then* $\delta(A^*) \in \mathcal{S}_p$ *and*

$$\|\delta(A^*)\|_p \leq 2c_p\|\delta(A)\|_p; \tag{4.15}$$

(ii) *if moreover* $p = 2$ *and* δ *is closed then*

$$\|\delta(f(A))\|_2 \leq k(f)\|\delta(A)\|_2 \tag{4.16}$$

for each Lipschitz function f *on* $\sigma(A)$ *with Lipschitz constant* $k(f)$.

The extension of part (ii) to all $p \in (1, \infty)$ was obtained in [13].

Let $A \in \mathcal{B}(H)$ be a normal operator. As it was mentioned above the map $AX - XA \mapsto A^*X - XA^*$ is not bounded in general as a map on $\mathcal{B}(H)$. However this map and more general maps of the form $AX - XA \mapsto f(A)X - Xf(A)$ (f is a measurable function) are norm closable. In fact, if $AX_n - X_nA \to 0$ and $f(A)X_n - X_nf(A) \to B$ as $n \to \infty$ then

$$[B, A] = \lim_{n\to\infty}[[f(A), X_n], A] = \lim_{n\to\infty}[f(A), [X_n, A]] = 0.$$

Thus B is in the commutant $\widetilde{\{A\}'}$ of A. Hence B commutes with $f(A)$ and therefore belongs to $\overline{\delta_{f(A)}(\mathcal{B}(H))} \cap \ker \delta_{f(A)}$ which is $\{0\}$ by [2].

It would be interesting to see what happen if we weaken the condition on the topology replacing it by the weak-*-topology on $\mathcal{B}(H)$. It was proved in [21] that the Fuglede map $AX - XA \mapsto A^*X - XA^*$ is not w^*-closable if $\sigma(A)$ has non-empty interior and the spectral measure of A is equivalent to the Lebesgue measure on the interior of $\sigma(A)$. The proof uses a characterization of so-called Toeplitz w^*-closable multipliers given in [21]. However the method does not work for general maps of the type $AX - XA \mapsto f(A)X - Xf(A)$. Some sufficient conditions are established in [21].

Problem 8. For which continuous functions f and normal operators $A \in \mathcal{B}(H)$ is the map on $\mathcal{B}(H)$ given by $AX - XA \mapsto f(A)X - Xf(A)$ closable in weak-*-topology?

References

[1] A. Abdessemed, E.B. Davies, *Some commutator estimates in the Schatten classes*, J. London Math. Soc. (2) 39 (1989), no. 2, 299–308.

[2] J. Anderson and C. Foias, *On properties which normal operators share with normal derivations and related operators*, Pac. J. Math. 61 (1975), 313–325.

[3] J. Anderson, *Derivation ranges and the identity*, Proc. Amer. Math. Soc. 38 (1973), 135–140.

[4] C.A. Berger B.I. Shaw, *Selfcommutators of multicyclic hyponormal operators are always trace class*, Bull. Amer. Math. Soc. 79 (1973), no. 6, 1193–1199.

[5] K. Boyadziev,*Commuting C_0-groups and the Fuglede–Putnam theorem*, Studia Math. 81 (1985), no. 3, 303–306.

[6] R. Curto and L. Fialkow, *The spectral picture of $[L_A, R_B]$*, J. Funct. Anal. 71 (1987), 371–392.

[7] M.R. Duggal, *On Fuglede-Weiss theorem*, Indiana Univ. Math. J. 36 (1987), 526–534.

[8] E.A. Gorin, *A generalization of a theorem of Fuglede*, Algebra and Analysis, 5 (1993), 83–97.

[9] E.A. Gorin, M.I. Karahanyan, *Asymptotic version of Fuglede–Putnam Theorem on commutators for elements of Banach algebras*, Mathem. Zametki 22 (1977), no. 2, 179–188.

[10] Yu.B. Farforovskaya, *Example of a Lipschitz function of selfadjoint operators that gives a non-nuclear increment under a nuclear perturbation*, J. Soviet Math. 4 (1975), 426–433.

[11] Y. Ho, *Commutants and derivation ranges*, Tohoku Math. J. 27 (1975), 509–514.

[12] E. Kissin and V.S. Shulman, *Classes of operator-smooth functions. I. Operator-Lipschitz functions*, Proc. Edinburgh Math. Soc. 48 (2005), 151–173.

[13] E. Kissin, D. Potapov, V. Shulman and F. Sukochev, *Lipschitz functions, symmetrically normed ideals and unbounded derivations*, in preparation.

[14] F. Kittaneh, *On Lipschitz functions of normal operators*, Proc. Amer. Math. Soc. 94 (1985), 416–418.

[15] D. Potapov and F. Sukochev, *Operator-Lipschitz functions in Schatten-von Neumann classes,* Acta Math., to appear.

[16] C. Putnam, *On normal operators on Hilbert space*, Amer. J. Math. 73 (1951), 357–362.

[17] M. Rosenblum, *On a theorem of Fuglede and Putnam*, J. London Math. Soc. 33 (1958), 376–377.

[18] V.S. Shulman, *Linear operator equations with generalized scalar coefficients*, (Russian) Dokl. Akad. Nauk SSSR 225 (1975), no. 1, 56–58.

[19] V.S. Shulman, *Multiplication operators and spectral synthesis*, Doclady AN SSSR 313 (1990), no. 5, 1047–1051.

[20] V.S. Shulman, *Some Remarks on the Fuglede-Weiss Theorem*, Bull. London Math. Soc. 28 (1996), 385–392.

[21] V.S. Shulman, I.G. Todorov and L. Turowska, *Closable multipliers*, to appear in Integral Equations Operator Theory, arXiv: 1001.4638.

[22] V. Shulman and L. Turowska, *Operator synthesis II: Individual synthesis and linear operator equations* J. Reine Angew. Math. 590 (2006), 143–187.

[23] V. Shulman and L. Turowska, *Beurling-Pollard type theorems*, J. Lond. Math. Soc. (2) 75 (2007), no. 2, 330–342.

[24] A. Turnšek, *On the range of elementary operators*, Publ. Math. Debrecen 63 (2003), no. 3, 293–304.

[25] D. Voiculescu, *Remarks on Hilbert-Schmidt perturbation of almost-normal operators. Topics in modern operator theory (Timişoara/Herculane,* 1980), pp. 311–318, Operator Theory: Adv. Appl., 2, Birkhäuser, Basel-Boston, Mass., 1981.

[26] D. Voiculescu, *Some extensions of quasitriangularity*, Rev. Roumaine Math. Pures Appl. 18 (1973), 1303–1320.

[27] Gary Weiss, *The Fuglede commutativity theorem modulo the Hilbert–Schmidt class and generating functions for matrix operators. I*, Trans. Amer. Math. Soc. 246 (1978), 193–209.

[28] Gary Weiss, *An extension of the Fuglede–Putnam theorem modulo the Hilbert–Schmidt class to operators of the form* $\sum M_n X N_n$ Trans. Amer. Math. Soc. 278 (1983), no. 1, 1–20.

[29] Gary Weiss, *The Fuglede commutativity theorem modulo the Hilbert–Schmidt class and generating functions for matrix operators. II*, J. Operator Theory 5 (1981), no. 1, 3–16.

V.S. Shulman
Department of Mathematics, Vologda State Technical University, Vologda, Russia
e-mail: `shulman_v@yahoo.com`

L. Turowska
Department of Mathematical Sciences, Chalmers University of Technology and the University of Gothenburg, Gothenburg SE-412 96, Sweden
e-mail: `turowska@chalmers.se`

Computation Versus Formulae for Norms of Elementary Operators

Richard M. Timoney

Abstract. A long-standing problem was to find a formula for the norm of an elementary operator acting on a C^*-algebra (or a matrix algebra). As with the best problems, the problem was quick to explain, although there was no conjectured answer. The solution which has been found involves concepts that are almost equally simple to explain, but the effective use of the formula is more subtle. It does lead to insights into the structure of the algebras. The techniques involved in justifying the formula are useful in practical and theoretical ways.

Mathematics Subject Classification (2000). Primary 47B47; Secondary 46L07, 46L05.

Keywords. Completely bounded norm; matrix numerical range; tracial geometric mean; primal ideal; Glimm ideal; factorial state; pure state; central Haagerup tensor product.

1. Preamble

Elementary operators are linear operators on algebras which are defined in an appealingly simple algebraic way. In view of the very significant amount of research on them, and the results that have been obtained in different directions, the term 'elementary' may be a bit of a misnomer. Our aim here is to give a flavour of some of the work on the topic, but to concentrate on results about norms of elementary operators acting on C^*-algebras, the simplest examples being the algebras $M_n = M_n(\mathbb{C})$ of $n \times n$ complex matrices.

Recall that, by definition, an elementary operator on a unital algebra A is a map $T\colon A \to A$ which is expressible as a finite sum of two-sided multiplication maps $x \mapsto axb$. We will write $\mathcal{E}\ell(A)$ for the elementary operators on A. Thus each

R.E. Curto and M. Mathieu (eds.), *Elementary Operators and Their Applications*, Operator Theory: Advances and Applications 212, DOI 10.1007/978-3-0348-0037-2_7,

$T \in \mathcal{E}\ell(A)$ is expressible as

$$Tx = \sum_{j=1}^{\ell} a_j x b_j \tag{1.1}$$

where $\ell \geq 1$ and a_j and b_j belong to A. (We should allow the trivial case $T = 0$ and $\ell = 0$, but it is also included with $\ell = 1$, $a_1 = b_1 = 0$.) Left multiplication operators $L_a(x) = ax$ (with $a \in A$) and right multiplications $R_b(x) = xb$ (with $b \in A$) are included, and perhaps this is already an indication that the whole structure of the algebra is somehow reflected by properties of the elementary operators on it.

We will explain substantive results that connect the structure of an algebra with the elementary operators on it, but there are many more results that explain the interest in these operators. Some results may not even refer to the terminology, such as results that all derivations on certain algebras are inner, which have cohomological interpretations. In quantum information theory the term 'superoperator' is used for operators on algebras. At least for $A = M_n$, these are all elementary operators.

For unital Banach algebras A, the terminology 'elementary operator' seems to have been coined by Lumer and Rosenblum [21], who required that the sets $\{a_j : 1 \leq j \leq \ell\}$ and $\{b_j : 1 \leq j \leq \ell\}$ separately commute. Their concerns were largely with the spectrum of T, and they were interested in generalising Sylvester's theorem concerning solutions x of matrix equations $(L_a - R_b)(x) = y$. We refer to the article [9] for a wide-ranging survey of the consequences and applications of Sylvester's work. The work of Fong and Sourour [14] is an early substantial contribution where the commutativity of the sets $\{a_j\}$ and $\{b_j\}$ is not required. The first workshop in the series, held in 1991 (proceedings volume [25]) reflected substantial progress during the preceding decade, and the second was held at the University of Helsinki, Finland during September, 2001. Rather than attempting to survey the different developments considered by various workers (as is done in [28]), we concentrate here on a rather specific strand aimed at finding the norm of T as in (1.1) for the case of a C^*-algebra A. This problem was stated explicitly in [25] and there is a comprehensive survey by Mathieu [26] of progress in the previous decade. Perhaps the problem can be considered to have been solved recently.

Except in the last two sections, A will always be a unital C^*-algebra. When we refer to an ideal, we always mean a closed two-sided ideal.

2. Observations

It is clear that if $I \subset A$ is an ideal and $T \in \mathcal{E}\ell(A)$, then $T(I) \subset I$. Moreover, with T as in (1.1), T induces $T^I \in \mathcal{E}\ell(A/I)$ via $T^I(x + I) = \sum_{j=1}^{\ell} a_j x b_j + I = Tx + I$.

It seems that the property of preserving ideals of A, which is almost tautological, underlies many of the deeper results linking properties of elementary operators to the ideal structure of A. Magajna [22, Corollary 2.3] showed that in the (relatively weak) sense of point norm convergence, all bounded linear operators on A that preserve ideals are limits of elementary operators. (There are characterisations of C^*-algebras A where stronger approximation results hold in [24, 23].)

We use $\mathcal{B}(H)$ for the bounded linear operators on a Hilbert space H, and write $\hat{A}$ for the unitary equivalence classes $[\pi]$ of irreducible representations $\pi\colon A \to \mathcal{B}(H_\pi)$. For $T \in \mathcal{E}\ell(A)$ as in (1.1) and $\pi \in \hat{A}$, we can define $T^{\ker \pi} \in \mathcal{E}\ell(A/\ker \pi)$ using the ideal $I = \ker \pi$. Essentially equivalently we can use the natural isomorphism $A/\ker \pi \cong \pi(A)$ to treat this as an elementary operator on $\pi(A)$. Instead we will use the operator $T^\pi \in \mathcal{E}\ell(\mathcal{B}(H_\pi))$ which is given by

$$T^\pi(y) = \sum_{j=1}^{\ell} \pi(a_j) y \pi(b_j) \qquad (y \in \mathcal{B}(H_\pi)),$$

Since T^π is continuous in the weak* (or ultraweak) topology of $\mathcal{B}(H_\pi)$, T^π restricts to the canonical element of $\mathcal{E}\ell(\pi(A))$ associated with T, and $\pi(A)$ is weak*-dense in $\mathcal{B}(H_\pi)$, we can see that T^π is independent of the representation (1.1) of $T \in \mathcal{E}\ell(A)$. From the Kaplansky density theorem $\|T^\pi\|$ is the same as the norm of its restriction to $\pi(A)$ and then faithfulness of the reduced atomic decomposition implies that

$$\|T\| = \sup_{\pi \in \hat{A}} \|T^\pi\|. \tag{2.1}$$

See [1, Theorem 5.3.12] for a more general result than (2.1).

We can see from (2.1) that we could calculate $\|T\|$ for $T \in \mathcal{E}\ell(A)$ if we knew how to deal with the case $A = \mathcal{B}(H)$. We will first consider a solution by Stampfli [30] for the special case of inner derivations.

3. Inner derivations on $A = \mathcal{B}(H)$

For $c \in A$, the inner derivation $\delta_c \in \mathcal{E}\ell(A)$ is given by $\delta_c(x) = xc - cx$. It is probably not necessary to mention that derivations in general and inner derivations in particular play an important role in many ways. Their connection to Lie theory and the Leibniz rule they satisfy partly explains the interest in them. An obvious estimate is that $\|\delta_c\| \leq 2\|c\|$. Since $\delta_{c-\lambda} = \delta_c$ we have $\|\delta_c\| \leq 2\inf_{\lambda\in\mathbb{C}} \|c - \lambda\|$ (where $c - \lambda$ means $c - \lambda 1_A$ with $1_A \in A$ the identity).

Theorem 3.1 (Stampfli [30]). *For $A = \mathcal{B}(H)$, $\|\delta_c\| = 2\inf_{\lambda\in\mathbb{C}} \|c - \lambda\|$.*

We contend that the importance of this result stems not just from the elegance of the formula, but also from the method of proof. The proof used a modification of the numerical range of $c - \lambda$ to characterise the $\lambda \in \mathbb{C}$ where the infimum is attained. This characterisation implies an algorithm for checking any guess one might be able to make about the infimum.

Recall that the usual (spatial) numerical range of $c \in \mathcal{B}(H)$ is

$$W(c) = \{\langle c\xi, \xi\rangle : \xi \in H, \|\xi\| = 1\},$$

where $\langle\cdot,\cdot\rangle$ denotes the inner product on H (which we take to be linear in the first variable). Standard facts about $W(c)$ are that it is a convex subset of $\mathbb{C}$ (the Toeplitz–Hausdorff theorem), it need not be closed (unless $\dim H < \infty$), the closure $\overline{W(c)}$ contains the spectrum of c, and $\overline{W(c)}$ coincides with the convex hull of the spectrum of c if c is normal ($c^*c = cc^*$). (See [18] for instance.)

Stampfli [30] used a subset $W_0(c) \subset \overline{W(c)}$ which can be conveniently explained using the joint numerical range of (c^*c, c). For a tuple $(c_1, c_2, \ldots, c_n)$ of operators on H, their joint numerical range is

$$W(c_1, c_2, \ldots, c_n) = \{(\langle c_j\xi, \xi\rangle)_{j=1}^n : \xi \in H, \|\xi\| = 1\} \subset \mathbb{C}^n.$$

Using the closure of $W(c^*c, c)$, we can describe

$$W_0(c) = \{z \in \mathbb{C} : (\|c\|^2, z) \in \overline{W(c^*c, c)}\}.$$

Stampfli [30] showed that (for $c \in \mathcal{B}(H)$)

$$\|\delta_c\| = 2\|c - \lambda\| \iff 0 \in W_0(c - \lambda). \tag{3.1}$$

We cannot express $W_0(c-\lambda)$ in terms of $W_0(c)$ and λ, but a very simple calculation reveals that

$$W((c-\lambda)^*(c-\lambda), c-\lambda) = \{(x + |\lambda|^2 - 2\Re(\bar{z}\lambda), z - \lambda) : (x, z) \in W(c^*c, c)\}.$$

Thus $W_0(c-\lambda)$ must be computable from the joint numerical range $W(c^*c, c)$. Also the norm $\|c\|$ is given by $\|c\|^2 = \sup\{x : (x, z) \in W(c^*c, c) \text{ for some } z \in \mathbb{C}\}$, and $W(c^*c, c) \subset \mathbb{R} \times \mathbb{C}$ contains all the information needed to identify $\|\delta_c\|$.

Our aim here is to explain a generalisation of this strategy that applies to $T \in \mathcal{E}\ell(\mathcal{B}(H))$. We should point out that we are omitting discussion of many developments arising from Stampfli's paper relating to $\|\delta_c\|$ for $c \in A$, A more general than $\mathcal{B}(H)$. However, those aspects are rather well summarised in [1, §4.1, 4.6].

4. Haagerup estimates

Consider now $T \in \mathcal{E}\ell(\mathcal{B}(H))$ given as in (1.1) with $a_j, b_j \in \mathcal{B}(H)$ $(1 \le j \le \ell)$. Following an idea of Haagerup [17], we express Tx as a 1×1 matrix with entry in $\mathcal{B}(H)$ and as a (block) matrix product

$$Tx = \begin{bmatrix} a_1 & a_2 & \ldots & a_\ell \end{bmatrix} \begin{bmatrix} x & 0 & \cdots & 0 \\ 0 & x & & 0 \\ \vdots & & \ddots & \\ 0 & 0 & \cdots & x \end{bmatrix} \begin{bmatrix} b_1 \\ b_2 \\ \vdots \\ b_\ell \end{bmatrix} = \mathbf{a}x^{(\ell)}\mathbf{b},$$

where $\mathbf{a}$ denotes the row, $x^{(\ell)}$ denotes the block diagonal with ℓ copies of x on the diagonal, and $\mathbf{b}$ denotes the column. Treating this as a factorisation of Tx through

$H \to H^\ell \to H^\ell \to H$ (with H^ℓ the Hilbert space direct sum of ℓ copies of H), we get

$$\|Tx\| \leq \|\mathbf{a}\|\|\mathbf{b}\|\|x\| = \sqrt{\left\|\sum_{j=1}^{\ell} a_j a_j^*\right\| \left\|\sum_{j=1}^{\ell} b_j^* b_j\right\|}\, \|x\|.$$

An important part of Haagerup's contribution in [17] was that the resulting estimate for $\|T\|$ is also an estimate for potentially bigger norms associated with T. We can identify $k \times k$ block matrices $X = (x_{r,s})_{r,s=1}^k$ with entries $x_{r,s} \in \mathcal{B}(H)$, that is the elements of $M_k(\mathcal{B}(H))$, with operators on H^k by treating vectors in H^k as columns ($k \times 1$ matrices with entries in H). In summary, we have $M_k(\mathcal{B}(H)) \cong \mathcal{B}(H^k)$. We associate with any linear operator $T\colon \mathcal{B}(H) \to \mathcal{B}(H)$ the operators $T^{(k)}\colon M_k(\mathcal{B}(H)) \to M_k(\mathcal{B}(H))$ given by

$$T^{(k)}(X) = T^{(k)}((x_{r,s})_{r,s=1}^k) = (Tx_{r,s})_{r,s=1}^k.$$

In tensor notation, we could regard $M_k(\mathcal{B}(H))$ as $\mathcal{B}(H) \otimes M_k$ and then $T^{(k)}$ becomes $T \otimes \mathrm{id}$ (where id means the identity on M_k). We define $\|T\|_k = \|T^{(k)}\|$. For T bounded, we do always have $\|T\|_k < \infty$ for each k. However, a considerable ingredient of modern operator space theory is based on the fact operators where $\sup_{k \geq 1} \|T\|_k < \infty$, known as *completely bounded* operators, have a special significance. The *completely bounded norm* (or CB-norm) of such T is

$$\|T\|_{cb} \stackrel{def}{=} \sup_{k \geq 1} \|T\|_k.$$

For $T \in \mathcal{E}\ell(\mathcal{B}(H))$ as in (1.1), it is easy to see that $T^{(k)} \in \mathcal{E}\ell(M_k(\mathcal{B}(H)))$ (for each $k \geq 1$) because we can write

$$T^{(k)}(X) = \sum_{j=1}^{\ell} a_j^{(k)} X b_j^{(k)} \tag{4.1}$$

(using the same notation for $a_j^{(k)}$ and $b_j^{(k)}$ as we used above for the block diagonal $x^{(\ell)}$). If we now use the block matrix trick again, with T replaced now by $T^{(k)}$, we get $\|T\|_k \leq \|\mathbf{a}\|\|\mathbf{b}\|$, an estimate independent of k and hence a bound for $\|T\|_{cb}$.

There is a further avenue to improve the estimate. Though we begin with T expressed in the form (1.1), there are many such expressions giving T. We may consider the tensor

$$u = \sum_{j=1}^{\ell} a_j \otimes b_j \in \mathcal{B}(H) \otimes \mathcal{B}(H)$$

and then there is a well-defined linear map $\theta\colon \mathcal{B}(H) \otimes \mathcal{B}(H) \to \mathcal{E}\ell(\mathcal{B}(H))$ given by $\theta(u) = T$, or

$$\theta(u)(x) = \sum_{j=1}^{\ell} a_j x b_j \qquad (x \in \mathcal{B}(H)).$$

The tensor norm defined by Haagerup [17] is now known as the Haagerup tensor norm and is defined (on $\mathcal{B}(H) \otimes \mathcal{B}(H)$) by

$$\|u\|_h = \inf\left\{ \|\mathbf{a}\|\|\mathbf{b}\| : u = \sum_{j=1}^{\ell} a_j \otimes b_j \right\} = \inf\left\{ \sqrt{\left\|\sum_{j=1}^{\ell} a_j a_j^*\right\| \left\|\sum_{j=1}^{\ell} b_j^* b_j\right\|} \right\}.$$

The insight of Haagerup was that, not only do we get a bound on $\|T\|_{cb}$, we actually get equality.

Theorem 4.1 (Haagerup [17]). *For $u = \sum_{j=1}^{\ell} a_j \otimes b_j \in \mathcal{B}(H) \otimes \mathcal{B}(H)$ and $T = \theta(u) \in \mathcal{E}\ell(\mathcal{B}(H))$, we have $\|T\|_{cb} = \|u\|_h$.*

Haagerup's argument can be found in [1, §5.4]. It would be hard to overstate the influence of this contribution by Haagerup, but we will limit ourselves to some of its consequences for elementary operators. One of the early discoveries was that the infimum in the definition of $\|u\|_h$ is attained, and moreover that if $u = \sum_{j=1}^{\ell} \tilde{a}_j \otimes \tilde{b}_j$ then there is another expression $u = \sum_{j=1}^{\ell} a_j \otimes b_j$ where

$$\|u\|_h = \|\mathbf{a}\|\|\mathbf{b}\| = \frac{1}{2}(\|\mathbf{a}\|^2 + \|\mathbf{b}\|^2) \tag{4.2}$$

and $a_j \in \mathrm{span}(\{\tilde{a}_k : 1 \le k \le \ell\})$, $b_j \in \mathrm{span}(\{\tilde{b}_k : 1 \le k \le \ell\})$. This fact is usually summarised as 'injectivity' of the Haagerup tensor norm (in the context of operator spaces). There are by now at least 4 authoritative recent monographs on operator space theory, one of which is Effros and Ruan [12]. We refer to those for details missed out above.

5. Algorithms

While it is very useful to have Haagerup's Theorem and the fact that equality is attainable via expressions of $u \in \mathcal{B}(H) \otimes \mathcal{B}(H)$ within a specific finite-dimensional space, this does not necessarily provide a direct way to find $\|u\|_h$ in any given example. It seems necessary to search for lower bounds for $\|T\|_{cb} \ge \|T\|_k$ with arbitrarily large k and upper bounds of the form $\|\mathbf{a}\|\|\mathbf{b}\|$.

In [31] a number of results were established that help in this respect, via the use of (joint) numerical range considerations. These results were inspired by the considerations of Stampfli [30] outlined in §3. The joint numerical ranges are those of the ℓ^2-tuples $\{a_i a_j^* : 1 \le i, j \le \ell\}$ and $\{b_i^* b_j : 1 \le i, j \le \ell\}$ but it is convenient to consider them as subsets of the space of $\ell \times \ell$ matrices. We define the *matrix numerical range* $W_m(\mathbf{b})$ associated with the column $\mathbf{b}$ by

$$W_m(\mathbf{b}) = \{(\langle b_j^* b_i \xi, \xi\rangle)_{i,j=1}^{\ell} : \xi \in H, \|\xi\| = 1\}. \tag{5.1}$$

We will also consider the subset of the closure with maximal trace $\|\mathbf{b}\|^2$, which is $W_{m,e}(\mathbf{b}) = \{X \in \overline{W_m(\mathbf{b})} : \mathrm{trace}(X) = \|\mathbf{b}\|^2\}$. Similarly for the row $\mathbf{a}$ we apply adjoints to get a column $\mathbf{a}^*$ and consider $W_m(\mathbf{a}^*)$ and $W_{m,e}(\mathbf{a}^*) = \{Y \in \overline{W_m(\mathbf{a}^*)} : \mathrm{trace}(Y) = \|\mathbf{a}\|^2\}$.

Theorem 5.1 (Theorem 1). *For* $u = \sum_{i=1}^{\ell} a_i \otimes b_i \in \mathcal{B}(H) \otimes \mathcal{B}(H)$, $T = \theta(u)$, *we have*

$$\|T\| = \frac{1}{2}(\|\mathbf{a}\|^2 + \|\mathbf{b}\|^2) = \|u\|_h \iff W_{m,e}(\mathbf{a}^*) \cap W_{m,e}(\mathbf{b}) \neq \emptyset.$$

Recall that $\|T\| \leq \|T\|_{cb}$ and strict inequality is to be expected in general. Thus, although we can always rewrite the tensor u so that (4.2) holds (and also we have $\|u\|_h = \|T\|_{cb}$), this theorem cannot always be applied directly.

However, let us now recall (4.1). Write $\mathbf{a}^{(k)}$ for the row associated with (4.1) and $\mathbf{b}^{(k)}$ for the column. Then $W_m(\mathbf{b}^{(k)}) = \mathrm{co}_k(W_m(\mathbf{b})$ is easy to verify, where $\mathrm{co}_k(\cdot)$ means the set of all convex combinations of at most k elements of a set. Similarly $W_m((\mathbf{a}^{(k)})^*) = \mathrm{co}_k(W_m(\mathbf{a}^*)$. It is then a short step from Theorem 1 to get the following.

Theorem 5.2 (Theorem *k*). *For* $k \geq 1$ *and* u, T *as in Theorem* 1, *we have*

$$\|T\|_k = \frac{1}{2}(\|\mathbf{a}\|^2 + \|\mathbf{b}\|^2) = \|u\|_h \iff \mathrm{co}_k(W_{m,e}(\mathbf{a}^*)) \cap \mathrm{co}_k(W_{m,e}(\mathbf{b})) \neq \emptyset.$$

More or less by letting k become large, we get the following. Here $\mathrm{co}(\cdot)$ will denote the convex hull of a set.

Theorem 5.3 (Theorem CB). *Let* u *and* T *be as in Theorem* 1. *We have*

$$\frac{1}{2}(\|\mathbf{a}\|^2 + \|\mathbf{b}\|^2) = \|u\|_h \iff \mathrm{co}(W_{m,e}(\mathbf{a}^*)) \cap \mathrm{co}(W_{m,e}(\mathbf{b})) \neq \emptyset.$$

Moreover $\|T\|_k = \|T\|_{cb}$ *for* $k = \min(\ell, \dim H)$.

The proof that $k = \ell$ suffices for $\|T\|_k = \|T\|_{cb}$ uses a generalisation of the classical Toeplitz–Hausdorff theorem to tuples. It is not true that the joint numerical range W of more than one operator on H needs to be convex, but such numerical ranges satisfy $\mathrm{co}_k(W) = \mathrm{co}(W)$ for a smaller value of k than holds for arbitrary subsets of a vector spaces with the appropriate dimension. (ℓ^2 is the real dimension here. See [31, Remark 2.5].)

Starting with $u = \sum_{j=1}^{\ell} \tilde{a}_j \otimes \tilde{b}_j$, we know we can find another expression $u = \sum_{j=1}^{\ell} a_j \otimes b_j$ as in (4.2). We use $\tilde{\mathbf{a}}$ and $\tilde{\mathbf{b}}$ as shorthand for the row and column associated with the $\tilde{a}_j$ and $\tilde{b}_j$ (by analogy with $\mathbf{a}$ and $\mathbf{b}$). But then $W_m(\mathbf{a}^*)$ is expressible in terms of the matrix numerical range $W_m(\tilde{\mathbf{a}}^*)$. The norm $\|\mathbf{a}\|$ is also computable from $W_m(\mathbf{a}^*)$. Similarly $W_m(\mathbf{b})$ and $\|\mathbf{b}\|$ are expressible in terms of $W_m(\tilde{\mathbf{b}})$. So, at least in principle, $\|T\|_{cb} = \|u\|_h$ is computable in terms of the matrix (or joint) numerical ranges $W_m(\tilde{\mathbf{a}}^*)$ and $W_m(\tilde{\mathbf{b}})$.

6. Formula for the norm on $\mathcal{B}(H)$

This last fact does not yet allow us to claim that $\|T\|$ is computable in terms of these same joint numerical ranges. We now continue to deal here with $u = \sum_{j=1}^{\ell} a_j \otimes b_i \in \mathcal{B}(H) \otimes \mathcal{B}(H)$ and the associated $T \in \mathcal{E\ell}(\mathcal{B}(H))$ given by (1.1).

A device used in [32] was to write $\|T\|$ in terms of norms of 'rank one' elementary operators, where the norm and CB norm coincide. Hence the norms of these rank one elementary operators are accessible via the numerical range techniques discussed above.

The idea is that

$$\|T\| = \sup\{\|x \mapsto \langle (Tx)\eta, \xi\rangle\| : \xi, \eta \in H, \|\xi\| = \|\eta\| = 1\},$$

as one can easily verify. This can be reformulated

$$\|T\| = \sup\{\|x \mapsto p_1(Tx)p_2\| : p_i^2 = p_i = p_i^* \in \mathcal{B}(H) \text{ of rank } = 1 (i = 1, 2)\}.$$

Using Theorem 1, it is possible to compute the norm of $x \mapsto p_1(Tx)p_2$ explicitly. The result is expressible in terms of a quantity called in [32] the *tracial geometric mean* of two positive semidefinite $\ell \times \ell$ matrices X and Y. It is given by

$$\operatorname{tgm}(X, Y) = \operatorname{trace}((X^{1/2}YX^{1/2})^{1/2}), \tag{6.1}$$

where the exponent $\frac{1}{2}$ means the positive semidefinite square root of the matrices. Here is the result. (We also use the notation (5.1) again).

Theorem 6.1 (Formula 1 [32]). *For* $u = \sum_{i=1}^{\ell} a_i \otimes b_i \in \mathcal{B}(H) \otimes \mathcal{B}(H)$, $T = \theta(u)$, *we have*

$$\|T\| = \sup\{\operatorname{tgm}(X, Y) : X \in W_m(\mathbf{a}^*), Y \in W_m(\mathbf{b})\}.$$

Using the same techniques that led from Theorem 1 to Theorem k above, we get the following.

Theorem 6.2 (Formula k [32]). *For* $k \geq 1$ *and* u, T *as in Formula* 1, *we have*

$$\|T\|_k = \sup\{\operatorname{tgm}(X, Y) : X \in \operatorname{co}_k(W_m(\mathbf{a}^*)), Y \in \operatorname{co}_k(W_m(\mathbf{b}))\}.$$

Theorem 6.3 (Formula CB [32]). *For* u, T *as in Formula* 1, *we have*

$$\|T\|_{cb} = \|u\|_h = \sup\{\operatorname{tgm}(X, Y) : X \in \operatorname{co}(W_m(\mathbf{a}^*)), Y \in \operatorname{co}(W_m(\mathbf{b}))\}.$$

Example. We might then be tempted to try Formula 1 on inner derivations δ_c, $c \in \mathcal{B}(H)$. The result is

$$\|\delta_c\| = \sup \operatorname{tgm}\left(\begin{bmatrix} 1 & -\langle c^*\xi, \xi\rangle \\ -\langle c\xi, \xi\rangle & \langle cc^*\xi, \xi\rangle \end{bmatrix}, \begin{bmatrix} \langle c^*c\eta, \eta\rangle & \langle c\eta, \eta\rangle \\ \langle c^*\eta, \eta\rangle & 1 \end{bmatrix}\right),$$

(with the supremum over unit vectors $\xi, \eta \in H$) but this does not seem to allow any easy way to "re-discover" Stampfli's formula from §3 above.

Remark 6.4. Although it was not noticed by the author in [32], the quantity $\operatorname{tgm}(X, Y)$ has been used extensively under the name *fidelity* in the literature on quantum information theory (at least when X and Y have trace 1). A source is [27, §9.2.2] (and references therein).

One might hope that the different perspectives of operator space theory (which is where the Haagerup tensor norm has been of fundamental importance) and quantum information theory would lead to new results in one area inspired by the other. Perhaps [33] gives some evidence for this possibility.

7. Formula for C^*-algebras

In view of (2.1), it is not difficult to generalise Formula 1 and Formula k to the case of $T \in \mathcal{E}\ell(A)$, for general (unital) C^*-algebras A. It helps to think in the language of states. On $\mathcal{B}(H)$, for $\xi \in H$ a unit vector, we have the associated state ω_ξ given by $\omega_\xi(x) = \langle x\xi, \xi\rangle$ (for $x \in \mathcal{B}(H)$). Then ω_ξ is a pure state of $\mathcal{B}(H)$ (in fact a pure normal state).

We write $\mathbb{S}(A)$ for the states on a C^*-algebra A (the positive linear functionals of norm 1), and $\mathbb{P}(A)$ for the pure states of A (the extreme points of $\mathbb{S}(A)$). Given $\phi \in \mathbb{S}(A)$ and an ℓ-tuple $\mathbf{b} = (b_1, b_2, \ldots, b_\ell)$ of elements of A (which we think of as a column matrix having entries in A), we define

$$Q(\mathbf{b}, \phi) = (\phi(b_j^* b_i))_{i,j=1}^{\ell}.$$

Using this notation, when $A = \mathcal{B}(H)$, $W_m(\mathbf{b}) = \{Q(\mathbf{b}, \omega_\xi) : \xi \in H, \|\xi\| = 1\}$.

We recall that each $\phi \in \mathbb{P}(A)$ gives rise, via the GNS construction, to an irreducible representation $\pi_\phi \colon A \to \mathcal{B}(H_\phi)$ and there is a unit vector $\xi_\phi \in H_\phi$ with $\phi(x) = \omega_{\xi_\phi}(\pi_\phi(x)))$. Conversely, given $\pi \in \hat{A}$ and a unit vector $\xi \in H_\pi$ we can get $\phi \in \mathbb{P}(A)$ from $\phi(x) = \omega_\xi(\pi(x))$.

For $\phi, \psi \in \mathbb{P}(A)$ we write $\phi \sim \psi$ if $[\pi_\phi] = [\pi_\psi]$ in $\hat{A}$.

Theorem 7.1 (Formula A-1 [32]). *For A a unital C^*-algebra, $T \in \mathcal{E}\ell(A)$ given by (1.1), $\mathbf{a} = [a_1, a_2, \ldots, a_\ell]$ (a row) and $\mathbf{b} = [b_1, b_2, \ldots, b_\ell]^t$ (a column), we have*

$$\|T\| = \sup\{\mathrm{tgm}(Q(\mathbf{a}^*, \phi_1), Q(\mathbf{b}, \phi_2)) : \phi_1, \phi_2 \in \mathbb{P}(A), \phi_1 \sim \phi_2\}.$$

To state a version of Formula k that applies to general A, we need a generalisation of (2.1) (see [1, Theorem 5.3.12]);

$$\|T\|_k = \sup_{\pi \in \hat{A}} \|T^\pi\|_k \quad (k \geq 1), \text{ and } \|T\|_{cb} = \sup_{\pi \in \hat{A}} \|T^\pi\|_{cb}. \tag{7.1}$$

We also need some more notation. For $k \geq 1$, a (type I) *factorial state of degree* (at most) k *on* A is $\phi \in \mathbb{S}(A)$ that is expressible as a convex combination

$$\phi = \sum_{i=1}^{k} t_i \phi_i$$

with $t_i \geq 0$ (all i), $\sum_{i=1}^k t_i = 1$, $\phi_i \in \mathbb{P}(A)$ (all i) and $\phi_i \sim \phi_j$ $(1 \leq i, j \leq k)$. We write $\mathbb{F}_k(A)$ for these states. For each $\phi \in \mathbb{F}_k(A)$ as above, there is an associated element of $\hat{A}$ given by the class $[\pi_{\phi_1}]$ of the GNS representation for ϕ_1. We denote this associated element of $\hat{A}$ by $\beta(\phi)$.

Theorem 7.2 (Formula A-k [32]). *For $k \geq 1$ and T, $\mathbf{a}$, $\mathbf{b}$ as in Formula A-1, we have*

$$\|T\|_k = \sup\{\mathrm{tgm}(Q(\mathbf{a}^*, \phi_1), Q(\mathbf{b}, \phi_2)) : \phi_1, \phi_2 \in \mathbb{F}_k(A), \beta(\phi_1) = \beta(\phi_2)\}.$$

In short, Formula A-1 expresses $\|T\|$ in terms of pairs or pure states which are related, and so a variation on numerical range ideas. To pass to Formula A-k, we take convex combinations of (at most) k pure states, but all the states must 'belong' to the same irreducible representation.

By using a $k \geq \ell$, or taking states in $\mathbb{F}_f(A) = \bigcup_{k=1}^{\infty} \mathbb{F}_k(A)$, we get:

Theorem 7.3 (Formula A-CB [32]). *For T, $\mathbf{a}$, $\mathbf{b}$ as in Formula A-1, we have*

$$\|T\|_{cb} = \sup\{\operatorname{tgm}(Q(\mathbf{a}^*, \phi_1), Q(\mathbf{b}, \phi_2)) : \phi_1, \phi_2 \in \mathbb{F}_f(A), \beta(\phi_1) = \beta(\phi_2)\}.$$

Moreover $\|T\|_{cb} = \|T\|_\ell$ and if $\sup_{\pi \in \hat{A}} \dim H_\pi := n < \infty$ (that is if A is subhomogeneous), then $\|T\|_{cb} = \|T\|_k$ for $k = \min(n, \ell)$.

8. Applications

One question we could consider is this one:

Question 8.1. *Does Haagerup's theorem $\|T\|_{cb} = \|u\|_h$ hold for general C^*-algebras A, or for many C^*-algebras?*

Here we are taking $u = \sum_{j=1}^{\ell} a_j \otimes b_j \in A \otimes A$ and $T \in \mathcal{E}\ell(A)$ as in (1.1). One way to make sense of $\|u\|_h$ is to consider A as a (unital) C^*-subalgebra of $\mathcal{B}(H)$ for some H and then use the earlier definition of $\|u\|_h$ for $u \in \mathcal{B}(H) \otimes \mathcal{B}(H)$. This is essentially relying on the injectivity of $\|\cdot\|_h$.

The ideas outlined in §4 apply to $T \in \mathcal{E}\ell(A)$ and show that we always have the inequality $\|T\|_{cb} \leq \|u\|_h$.

Vector states of $\mathcal{B}(H)$ restrict to give states of A, but there is no reason to expect the restrictions to be in $\mathbb{P}(A)$ or $\mathbb{F}_f(A)$. However, $\mathbb{S}(A)$ is the weak*-closed convex hull of $\mathbb{P}(A)$ and so we can use Formula-CB (see [7, Lemma 2.1]) to show that

$$\|u\|_h = \sup\{\operatorname{tgm}(Q(\mathbf{a}^*, \phi_1), Q(\mathbf{b}, \phi_2)) : \phi_1, \phi_2 \in \operatorname{co}(\mathbb{P}(A))\}. \tag{8.1}$$

Comparing with Formula A-CB, we can see that strict inequality seems likely in $\|T\|_{cb} \leq \|u\|_h$, at least unless we can resolve the discrepancy between $\mathbb{F}_f(A)$ on the one hand and $\operatorname{co}(\mathbb{P}(A))$ on the other. In $\mathbb{F}_f(A)$ we have rather restricted scope to include convex combinations of pure states. We do not actually need equality in the sets of states considered, but rather approximate equality of the relevant values (on $a_j a_i^*$ and $b_j^* b_i$). Essentially, if we are to have $\|T\|_{cb} = \|u\|_h$ for all $T \in \mathcal{E}\ell(A)$ we seem to need $\mathbb{F}_f(A)$ to be weak*-dense in $\mathbb{S}(A)$, which is a rather rare property. The class of C^*-algebras where this property holds is known to coincide with the prime C^*-algebras. (See [2] and [8, Proposition 2.2].) Primeness of A is equivalent to $L_a R_b \neq 0$ (on A) for nonzero $a, b \in A$, but is usually stated as the property that two nonzero ideals of A have nonzero product. It is a result of Mathieu (see [1, Proposition 5.4.11]) that $\|T\|_{cb} = \|u\|_h$ holds for all $u \in A \otimes A$ if and only if A is a prime C^*-algebra.

As yet, we cannot deduce this result of Mathieu from our remarks above because of the need to deal with pairs of states simultaneously in Formula A-CB.

However, this aspect turns out to be a technical issue, as the lemma below will imply.

In work of Archbold and Batty [3], the states $\phi \in \mathbb{S}(A)$ that are in the weak*-closure of $\mathbb{F}_f(A)$ were characterised as those where $\ker \pi_\phi$ is a primal ideal of A (π_ϕ being the associated GNS representation). An ideal $Q \subset A$ is called *n-primal* ($n \geq 2$) if whenever $J_1, J_2, \ldots, J_n$ are ideals of A with product $J_1 J_2 \ldots J_n = \{0\}$, then there is some $1 \leq j \leq n$ with $J_j \subset Q$. We call Q a *primal ideal* if it is n-primal for each n, and write Primal(A) for the set of primal ideals of A. Included among primal ideals are all primitive ideals (those of the form $\ker \pi$ with $\pi \in \hat{A}$). By a simple Zorn's lemma argument, every primal ideal contains a minimal primal ideal. We write Min-Primal(A) for the set of minimal primal ideals of A, Prim(A) for the primitive ideals.

Lemma 8.2 ([7, Lemma 2.2]). *Let $Q \subset A$ be a primal ideal and $\psi_1, \psi_2 \in \mathbb{S}(A/Q)$ (equivalently $\phi_1, \phi_2 \in \mathbb{S}(A)$ vanishing on Q). Then there exist commonly indexed nets $(\phi_{1,\alpha})_\alpha$ and $(\phi_{2,\alpha})_\alpha$ in $\mathbb{F}_f(A)$ so that $\beta(\phi_{1,\alpha}) = \beta(\phi_{2,\alpha})$ (all α), $\phi_{1,\alpha} \to \psi_1$ and $\phi_{2,\alpha} \to \psi_2$ (weak*-convergence).*

For $u = \sum_{j=1}^{\ell} a_j \otimes b_j \in A \otimes A$, and $I \subset A$ an ideal, we write $u^I = \sum_{j=1}^{\ell} (a_j + I) \otimes (b_j + I) \in (A/I) \otimes (A/I)$. It is easy to show that $\|u\|_h \geq \|u^I\|_h$ for any ideal I, and more generally that if $I \subset J \subset A$ are ideals then $\|u^I\|_h \geq \|u^J\|_h$. From our Formula A-CB, the Lemma and (8.1) (applied to u^Q) we can immediately deduce a result of Somerset.

Theorem 8.3 (Somerset [29]). *Let A be a (unital) C^*-algebra, $u = \sum_{j=1}^{\ell} a_j \otimes b_j \in A \otimes A$ and $T \in \mathcal{E}\ell(A)$ as in (1.1). Then*

$$\|T\|_{cb} = \sup\{\|u^Q\|_h : Q \in \text{Min-Primal}(A)\}.$$

Corollary 8.4 (Mathieu). *Let A be a (unital) C^*-algebra. Then A has the property that $\|T\|_{cb} = \|u\|_h$ for each $u = \sum_{j=1}^{\ell} a_j \otimes b_j \in A \otimes A$ and $T \in \mathcal{E}\ell(A)$ as in (1.1) if and only if A is prime.*

This corollary is evident from the theorem as the prime C^*-algebras are exactly those where $Q = \{0\}$ is a (minimal) primal ideal, and for the converse consider the definition given above for primeness. (It was established by Weaver [34] that there are [necessarily nonseparable] prime C^*-algebras which are not primitive. See also [10].)

To appreciate the result of Somerset, note that it is immediate from (7.1) and Haagerup's theorem that $\|T\|_{cb} = \sup\{\|u^P\|_h : P \in \text{Prim}(A)\}$. Here is an example to show how, in general, Min-Primal(A) can go further down the ideal structure of A than Prim(A) does.

Example. Let A be the C^*-subalgebra of $C([0,1], M_3)$, the continuous functions on the unit interval with values in M_3, given by restrictions on $x(0)$ and $x(1/n)$

$(n \geq 1)$ as follows:

$$x(0) = \begin{bmatrix} \lambda_1(x) & 0 & 0 \\ 0 & \lambda_2(x) & 0 \\ 0 & 0 & \lambda_3(x) \end{bmatrix}, \quad x\left(\frac{1}{n}\right) = \begin{bmatrix} \mu_n(x) & 0 \\ 0 & \theta_n(x) \end{bmatrix}.$$

Here $\lambda_j(x) \in \mathbb{C}$ $(1 \leq j \leq 3)$, $\mu_n(x) \in M_2$, $\theta_n(x) \in \mathbb{C}$. Then the minimal primal ideals of A are given by $Q_t = \{x \in A : x(t) = 0\}$ for $t \in [0,1]$. For $t = 0$ or $t = 1/n$, $Q_t \notin \mathrm{Prim}(A)$.

In view of the result in Corollary 8.4, we may realise that Question 8.1 was a somewhat naive question. In particular, prime C^*-algebras A have trivial center $Z(A)$ and we could observe that for $a, b \in A$, $z \in Z(A)$ the elementary operator T corresponding to $u = (az) \otimes b - a \otimes (zb)$ is $T = 0$. This leads us to consider the quotient of $A \otimes_h A$ by the closed linear span J_A of $\{(az) \otimes b - a \otimes (zb) : a, b \in A, z \in Z(A)\}$. We use the notation $A \otimes_{Z,h} A$ for the quotient $(A \otimes_h A)/J_A$ (with the quotient norm denoted $\|\cdot\|_{Z,h}$). We call $A \otimes_{Z,h} A$ the *central Haagerup tensor product* (see [1, §5.4]).

Question 8.5. *For which C^*-algebras A do we have $\|T\|_{cb} = \|u\|_{Z,h}$ for all $u = \sum_{j=1}^{\ell} a_j \otimes b_j \in A \otimes A$ and $T \in \mathcal{E}\ell(A)$ as in* (1.1).

This question has a satisfying answer in terms of the Glimm ideals of A, which are the ideals I of A such that $I \cap Z(A)$ is a maximal ideal in $Z(A)$ and generates I as an ideal of A. Somerset [29] showed that $\|u\|_{Z,h} = \sup \|u^G\|_h$, where the supremum is over all Glimm ideals G of A. It is known that every $Q \in \text{Min-Primal}(A)$ contains a unique Glimm ideal, and then from Theorem 8.3 it is clear that a sufficient condition (for a positive answer) is that Min-Primal(A) coincides with the set of Glimm ideals of A. In [5], this was shown to be necessary as well.

We quote now a result from [5, Theorem 17], which can be established using similar ideas but requires more detailed considerations. By θ_Z we mean the map from $A \otimes_{Z,h} A$ to $\mathrm{CB}(A)$ (the completely bounded linear operators on A with the CB norm) induced by $\theta\colon A \otimes A \to \mathcal{E}\ell(A)$ via continuous extension to the completion $A \otimes_h A$ and then passing to the quotient $A \otimes_{Z,h} A = (A \otimes_h A)/J_A$.

Theorem 8.6. *Let A be a unital C^*-algebra. Fix $\ell \geq 1$. Then*

$$\|\theta_Z(u)\|_{cb} = \|u\|_{Z,h}$$

holds for each $u = \sum_{j=1}^{\ell} a_j \otimes b_j \in A \otimes A$ if and only if every Glimm ideal in A is (ℓ^2+1)-primal.

Our aim was to show some examples where Formula A-CB suggest results that are not really obvious. There are other examples in [32, 7, 6] where considerations with the various formulae from §7 lead to new results that turn out to be best possible in various ways. We indicate some of these in the next section.

9. Generalisations for non-unital A

Most of the results have versions in the context of C^*-algebras that are not necessarily unital, but we have simplified our exposition by restricting to unital A above.

It is natural to consider the multiplier algebra $M(A)$ of A in the case when A has no identity. We recall that the multiplier algebra is a universal object associated with A, the largest algebra that contains A as an essential ideal. It is normally accepted that to define $\mathcal{E}\ell(A)$ in the non-unital case we should take those operators T on A expressible in the form (1.1) where $\ell \geq 1$ and $a_j, b_j \in M(A)$. There is then a map $\Theta\colon M(A) \otimes M(A) \to \mathcal{E}\ell(A)$ given by $\Theta\left(\sum_{j=1}^{\ell} a_j \otimes b_j\right) = T$ and an associated inequality $\|\Theta(u)\|_{cb} \leq \|u\|_h$. We should then try to recast our results from the unital case to this more general setting, a setting that has a number of advantages.

One obvious advantage is that it does allow left and right multiplication operators L_a and R_a, and inner derivations $\delta_a = R_a - L_a$, (for $a \in A$ or $a \in M(A)$) to be considered as genuine elementary operators.

A rather basic fact is that if $I \subset A$ is an ideal in the C^*-algebra A and $T \in \mathcal{E}\ell(A)$, then $T(I) \subset I$. [See [11, Proposition 1.8.5] for the fact that an ideal of an ideal of a C^*-algebra is an ideal.] The multiplier algebra $M(A)$ can be characterised (see [1, Chapter 1], for instance) by abstracting the properties of the pair $(L_m|_A, R_m|_A)$ of operators on A (for m an element of a C^*-algebra containing A as an essential ideal). It follows easily that for an ideal $I \subset A$ and $a, b \in M(A)$, the restriction to I of the map $x \mapsto L_a R_b x$ is in $\mathcal{E}\ell(I)$. More generally, $T \in \mathcal{E}\ell(A)$ implies that the restriction $T|_I \in \mathcal{E}\ell(I)$.

A fact which may be considered as reducing calculations to the unital case is that for $T = \Theta(u) \in \mathcal{E}\ell(A)$, $u \in M(A) \otimes M(A)$, we may also consider $T = \theta(u) \in \mathcal{E}\ell(M(A))$ and we have $\|\Theta(u)\|_k = \|\theta(u)\|_k$ (for each $k \geq 1$) by an application of Kaplansky's density theorem (see [5, Corollary 9]). This enables the formulae stated previously for unital A to be restated for general A at the expense of substituting $M(A)$ for A in many cases. For instance [5, Corollary 9] provides an answer to the analogue of Question 8.5, as follows.

Proposition 9.1. *Let A be a C^*-algebra. Then the equality $\|\Theta(u)\|_{cb} = \|u\|_{Z,h}$ holds for all $u \in M(A) \otimes M(A)$ (where $\Theta(u) \in \mathcal{E}\ell(A)$ and $\|\cdot\|_{Z,h}$ is the norm of $M(A) \otimes_{Z,h} M(A)$) if and only if all Glimm ideals of $M(A)$ are primal.*

A drawback of this statement, and of other statements we could deduce similarly, is that $M(A)$ can be quite a lot more complicated than A itself. In [5, Example 12], there is an example where A is quite well behaved in the sense that its primitive ideal space $\mathrm{Prim}(A)$ is compact Hausdorff, while $M(A)$ has Glimm ideals that are not 2-primal.

One way to view Glimm ideals and primal ideals (a way that works for both unital and non-unital A) is that they are related to the pathological nature of the topology of $\mathrm{Prim}(A)$. A definition equivalent to the one given earlier is that an ideal

I of A is n-primal when $P_1, P_2, \ldots, P_n \in \mathrm{Prim}(A/I)$ (which we identify with ideals in $\mathrm{Prim}(A)$ containing I) implies there exists a net $(P_\alpha)_\alpha$ in $\mathrm{Prim}(A)$ converging to all of $P_1, P_2, \ldots, P_n$ (see [3]). It follows that Min-Primal(A) and $\mathrm{Prim}(A)$ are different if and only if $\mathrm{Prim}(A)$ is not Hausdorff. A Glimm ideal is the kernel of an equivalence class in $\mathrm{Prim}(A)$, where $P, P' \in \mathrm{Prim}(A)$ are defined to be equivalent if $f(P) = f(P')$ for each bounded continuous function $f\colon \mathrm{Prim}(A) \to \mathbb{R}$. Thus, $\mathrm{Prim}(A)$ must be completely regular and Hausdorff for continuous functions on it to separate its points (and then $\mathrm{Glimm}(A)$ coincides with $\mathrm{Prim}(A)$).

What we learn from [5, Example 12] is that the pathology present in $M(A)$, or in $\mathrm{Prim}(M(A))$, may not be evident in $\mathrm{Prim}(A)$. Since results like Proposition 9.1 are necessary and sufficient conditions, we seem to have no way to state them in terms of conditions on A itself (without invoking $M(A)$).

It is worth pointing out that the formulae from §7 do in fact carry over to the non-unital case without change. The multiplier algebra enters in via the considerations that lead to (8.1), where we must substitute $\mathbb{P}(M(A))$ for $\mathbb{P}(A)$ when we take $u \in M(A) \otimes M(A)$.

We can do better in some sense if we stick to the original kind of elementary operator (1.1) with the restriction $a_j, b_j \in A$, even when A is not unital, thereby forsaking the advantages of allowing a_j and b_j to be in $M(A)$. This is the approach used in [7].

One hurdle is the selection of an appropriate definition of the central Haagerup tensor product $A \otimes_{Z,h} A$ for A not unital. Here we do still invoke the multiplier algebra, or more precisely its centre $Z(M(A))$, and consider the closed linear span J_A in $A \otimes_h A$ of $\{(az) \otimes b - a \otimes (zb) : a, b \in A, z \in Z(M(A))\}$. Note however that $Z(M(A))$ is directly related to $\mathrm{Prim}(A)$ by the Dauns-Hoffman theorem. Then defining $A \otimes_{Z,h} A = (A \otimes_h A)/J_A$ (a deviation from the notation of [1, Definition 5.4.14]) we do get a map $\theta_Z\colon A \otimes_{Z,h} A \to \mathrm{CB}(A)$ induced naturally from θ, a direct generalisation of Theorem 8.3 (to allow non-unital A) is given by [7, Theorem 3.5], and [7, Theorem 3.9] generalises Theorem 8.6 above (to allow non-unital A). We quote [7, Theorem 3.9] next.

Theorem 9.2. *Let A be a C^*-algebra and $\theta_Z\colon A \otimes_{Z,h} A \to \mathrm{CB}(A)$ the linear contraction given by $\theta_Z(a \otimes b + J_A)(x) = axb$. Then*

(i) *θ_Z is injective if and only if each $G \in \mathrm{Glimm}(A)$ is a 2-primal ideal of A.*
(ii) *θ_Z is an isometry if and only if each $G \in \mathrm{Glimm}(A)$ is a primal ideal of A.*

The main goal in [7] was to investigate conditions on A less general than 2-primality of all Glimm ideals, but more general than primality of all Glimm ideals, under which we can conclude that θ_Z is an isomorphism onto its range. This means conditions under which $L(A) := \|\theta_Z^{-1}\| < \infty$ (assuming, as we will do below, that θ_Z is injective). The results are more satisfactory in the case that A is stable, or by considering the map θ_Z not just on A but on all matrix algebras $M_n(A)$ at once.

Yet another way to characterise n-primality of an ideal I of A is to say that whenever $P_1, P_2, \ldots, P_n \in \mathrm{Prim}(A/I)$, then their intersection $\bigcap_{j=1}^{n} P_j$ is primal. If we exclude repetitions among the P_j and also exclude the possibility of containments $P_i \subsetneq P_j$, we can restate the condition. We say that an m-tuple $P_1, P_2, \ldots, P_m \in \mathrm{Prim}(A)$ is 'admissible' if there is no $i \neq j$ with $P_i \subseteq P_j$. Then n-primality of I is equivalent to $\bigcap_{j=1}^{m} P_j$ being primal for all admissible m-tuples $P_1, P_2, \ldots, P_m \in \mathrm{Prim}(A/I)$, for all $m \leq n$. In [7] we consider Glimm ideals G of A and for each admissible N-tuple $P_1, P_2, \ldots, P_N \in \mathrm{Prim}(A/G)$ we look at the associated collection of subsets $E \subset \{1, 2, \ldots, N\}$ such that $\bigcap_{j \in E} P_j$ is primal. These collections are abstract simplicial complexes (meaning collections of subsets containing all singleton subsets of the vertex set $\{1, 2, \ldots, N\}$ and closed under taking subsets; these are determined by their maximal members), and we say the simplicial complexes that arise in this way (for some $G \in \mathrm{Glimm}(A)$) are 'linked to the ideal structure of A'. When A is stable we show in [7, §6] that the isomorphism constant $L(A)$ depends only on the simplicial complexes linked to the ideal structure of A, hence only on the topological space $\mathrm{Prim}(A)$. There are just a few examples given in [7, §7] where we can compute $L(A)$, but these are sufficient to show that $L(A)$ can be infinite, any positive integer, or satisfy $1 < L(A) < 1 + \varepsilon$ (for any prescribed $\varepsilon > 0$). It remains open whether there are restrictions on the possible values of $L(A)$.

We refer to [7] for full details on these matters. Finally we mention that there are other applications of the formulae from §7 to be found in [6], where the notion of the upper multiplicity of an irreducible representation of a C^*-algebra is linked to considerations of norms of elementary operators.

10. Postscript

There have been some recent works (since the workshop) which are sufficiently pertinent to our overview above of earlier work as to deserve mention here.

I. Gogić [16, 15] has developed the work of B. Magajna [24, 23] cited in §2 above, which deals with understanding the closure in the CB norm (or in operator norm) of $\mathcal{E}\ell(A)$ (in the case of [23] the closure in the topology of pointwise norm convergence of $\{T \in \mathcal{E}\ell(A) : \|T\|_{cb} \leq 1\}$). Magajna considers the relationship of these closures with $\mathrm{ICB}(A) = \{T \in \mathrm{CB}(A) : T(I) \subseteq I$ for each ideal $I \subseteq A\}$. In [24], he characterized those separable A such that $\Theta(M(A) \otimes M(A))$ is dense in $\mathrm{ICB}(A)$ in CB norm (as those which are a finite direct sum of homogeneous C^*-algebras of finite type). On the other hand Gogić [16] showed that if A is separable and has $\Theta_Z(M(A) \otimes_{Z,h} M(A)) = \Theta(M(A) \otimes M(A))$ then A must be subhomogeneous and furthermore have a finite type property. Amongst other results, [15, §6] exhibits an example of a separable 2-subhomogeneous unital A such that $\mathcal{E}\ell(A)$ is complete in the operator norm, so that the range of $\Theta_Z = \theta_Z$ is just $\mathcal{E}\ell(A)$, moreover $\mathcal{E}\ell(A) \subsetneq \mathrm{ICB}(A)$, and there is a derivation δ of A that is in $\mathcal{E}\ell(A)$ but is not inner (*i.e.*, $\delta \neq \delta_c$ for $c \in A$). The main thrust in [15] is the relation between the range of θ_Z and derivations on A.

In private correspondence, I. Gogić has established the following result.

Proposition 10.1 (Gogić). *Assume A is an m-subhomogeneous C^*-algebra such that each Glimm ideal of A is 2-primal. Then $L(A) < \infty$ if and only if*

$$\sup\{|\operatorname{Prim}(A/G)| : G \in \operatorname{Glimm}(A)\} = N < \infty$$

(for $|\operatorname{Prim}(A/G)|$ the cardinality of $\operatorname{Prim}(A/G)$*).*

The justification is that $\hat{A}$ can be identified with $\operatorname{Prim}(A)$ (since A is liminal) and [13, Corollary 1 on p. 388] then implies that $|\operatorname{Prim}(A/Q)| \leq m$ for each $Q \in \operatorname{Primal}(A)$. Hence [7, Proposition 7.15] implies $L(A) \geq \sqrt{N/m}$, while [7, Corollary 7.11] (Proposition 9.1 for $N = 1$) gives $L(A) \leq \max(1, N - 1)$.

Using constructions inspired by generalising the simplicial spoke algebras of [7, 4.6] to infinitely many 'spokes' (or elaborating [15, §6]) together with Proposition 10.1, one can construct (unital separable) subhomogeneous C^*-algebras with all Glimm ideals 2-primal but $L(A) = \infty$ (and hence $\mathcal{E}\ell(A)$ not closed in CB norm, or in operator norm). Further properties considered in [15] can also be satisfied by such examples. (It is also possible to find subhomogeneous A to show that the hypothesis in Proposition 10.1 on 2-primality cannot be removed.)

A separate development is recent work of Archbold and Somerset [4] which sheds light on the relation between the primality properties of the Glimm ideals of $M(A)$ and properties of A itself. A striking result of [4] applies to separable C^*-algebras A and states that all Glimm ideals of $M(A)$ are primal (respectively n-primal for some $n \geq 2$) if and only if all Glimm ideals of A are both primal (respectively n-primal) and 'locally modular'. The definition of local modularity of $G \in \operatorname{Glimm}(A)$ is subtle. Let Y_G denote the boundary and U_G the interior of $\operatorname{Prim}(A/G)$ considered as a subset of $\operatorname{Prim}(A)$. G is called locally modular if for each $P \in Y_G$ there is an open neighbourhood V of $P \in \operatorname{Prim}(A)$ such that A/I is unital where $I = \bigcap\{R : R \in V \setminus U_G\}$. There are several other significant results in [4] relating the Glimm ideals of A with ideals of $M(A)$. Amongst the key ingredients used in [4] are recent results of A. Lazar [19, 20], which are expressed in topological terms but apply in particular to locally compact (not necessarily Hausdorff spaces) such as $\operatorname{Prim}(A)$. According to the formulation of primality in §9 (with nets), the topological interpretation of primality concerns (closed) limit sets. The considerations in [20] concern topologies on the set of closed limit sets of a locally compact space X and how they relate to the complete regularisation of X. An important result from [20] implies that two natural topologies on $\operatorname{Glimm}(A)$ (the *cr* topology being the weak topology induced by scalar-valued continuous functions on $\operatorname{Prim}(A)$, and the quotient topology from $\operatorname{Prim}(A)$) must coincide when A is σ-unital (in particular if A is separable). Delicate topological considerations are also a feature of the techniques used in [4].

Acknowledgment

The author would like to thank Martin Mathieu for constructive comments on an earlier draft of this paper.

References

[1] P. Ara and M. Mathieu, Local multipliers of C^*-algebras, Springer Monographs in Mathematics, Springer-Verlag London Ltd., London (2003).

[2] R.J. Archbold, *On factorial states of operator algebras*, J. Funct. Anal. **55** (1984), 25–38.

[3] R.J. Archbold and C.J.K. Batty, *On factorial states of operator algebras. III*, J. Operator Theory **15** (1986), 53–81.

[4] R.J. Archbold and D.W.B. Somerset, *Separation properties in the primitive ideal space of a multiplier algebra*, in preparation.

[5] R.J. Archbold, D.W.B. Somerset and R.M. Timoney, *On the central Haagerup tensor product and completely bounded mappings of a C^*-algebra*, J. Funct. Anal. **226** (2005), 406–428.

[6] R.J. Archbold, D.W.B. Somerset and R.M. Timoney, *Factorial states, upper multiplicity and norms of elementary operators*, J. Lond. Math. Soc. (2) **78** (2008), 707–722.

[7] R.J. Archbold, D.W.B. Somerset and R.M. Timoney, *Completely bounded mappings and simplicial complex structure in the primitive ideal space of a C^*-algebra*, Trans. Amer. Math. Soc. **361** (2009), 1397–1427.

[8] C.J.K. Batty and R.J. Archbold, *On factorial states of operator algebras. II*, J. Operator Theory **13** (1985), 131–142.

[9] R. Bhatia and P. Rosenthal, *How and why to solve the operator equation $AX-XB=Y$*, Bull. London Math. Soc. **29** (1997), 1–21.

[10] M.J. Crabb, *A new prime C^*-algebra that is not primitive*, J. Funct. Anal. **236** (2006), 630–633.

[11] J. Dixmier, Les C^*-algèbres et leurs représentations, Cahiers Scientifiques, Fasc. XXIX, Gauthier-Villars & Cie, Éditeur-Imprimeur, Paris (1964).

[12] E.G. Effros and Z.-J. Ruan, Operator spaces, *London Mathematical Society Monographs. New Series*, vol. 23, Oxford University Press, New York (2000).

[13] J.M.G. Fell, *The dual spaces of C^*-algebras*, Trans. Amer. Math. Soc. **94** (1960), 365–403.

[14] C.K. Fong and A.R. Sourour, *On the operator identity $\sum A_k X B_k \equiv 0$*, Canad. J. Math. **31** (1979), 845–857.

[15] I. Gogić, *Derivations which are inner as completely bounded maps*, Operators and Matrices **4** (2010), 193–211.

[16] I. Gogić, *Elementary operators and subhomogeneous C^*-algebras*, Proc. Edinb. Math. Soc. (2) (to appear).

[17] U. Haagerup, *The α-tensor product of C^*-algebras*, unpublished manuscript, Univ. of Odense (1980).

[18] P.R. Halmos, A Hilbert space problem book, *Graduate Texts in Mathematics*, vol. 19, Springer-Verlag, New York, second ed. (1982), encyclopedia of Mathematics and its Applications, 17.

[19] A.J. Lazar, *Hyperspaces of closed limit sets*, Methods Funct. Anal. Topology **16** (2010), 158–166.

[20] A.J. Lazar, *Quotient spaces determined by algebras of continuous functions*, preprint arXiv:0811.3280.

[21] G. Lumer and M. Rosenblum, *Linear operator equations*, Proc. Amer. Math. Soc. **10** (1959), 32–41.

[22] B. Magajna, *A transitivity theorem for algebras of elementary operators*, Proc. Amer. Math. Soc. **118** (1993), 119–127.

[23] B. Magajna, *Pointwise approximation by elementary complete contractions*, Proc. Amer. Math. Soc. **137** (2009), 2375–2385.

[24] B. Magajna, *Uniform approximation by elementary operators*, Proc. Edinb. Math. Soc. (2) **52** (2009), 731–749.

[25] M. Mathieu (ed.), Elementary operators and applications, World Scientific Publishing Co. Inc., River Edge, NJ (1992), in memory of Domingo A. Herrero.

[26] M. Mathieu, *The norm problem for elementary operators*, in Recent progress in functional analysis (Valencia, 2000), *North-Holland Math. Stud.*, vol. 189, North-Holland, Amsterdam (2001), pp. 363–368.

[27] M.A. Nielsen and I.L. Chuang, Quantum computation and quantum information, Cambridge University Press, Cambridge (2000).

[28] E. Saksman and H.-O. Tylli, *Multiplications and elementary operators in the Banach space setting*, in Methods in Banach space theory, *London Math. Soc. Lecture Note Ser.*, vol. 337, Cambridge Univ. Press, Cambridge (2006), pp. 253–292.

[29] D.W.B. Somerset, *The central Haagerup tensor product of a C^*-algebra*, J. Operator Theory **39** (1998), 113–121.

[30] J.G. Stampfli, *The norm of a derivation*, Pacific J. Math. **33** (1970), 737–747.

[31] R.M. Timoney, *Computing the norms of elementary operators*, Illinois J. Math. **47** (2003), 1207–1226.

[32] R.M. Timoney, *Some formulae for norms of elementary operators*, J. Operator Theory **57** (2007), 121–145.

[33] J. Watrous, *Distinguishing quantum operations having few Kraus operators*, Quantum Inf. Comput. **8** (2008), 819–833.

[34] N. Weaver, *A prime C^*-algebra that is not primitive*, J. Funct. Anal. **203** (2003), 356–361.

Richard M. Timoney
School of Mathematics
Trinity College
Dublin 2, Ireland
e-mail: richardt@maths.tcd.ie

Open Problems

This section contains problems that were either stated during the *Problem Session* at the workshop or arose in connection with the contributions to these proceedings. At the time of going into print, they appear to be open.

We start with some common notation. For a Banach algebra $\mathcal{A}$, let $\mathcal{E}\ell(\mathcal{A})$ be the algebra of elementary operators on $\mathcal{A}$. If $a, b \in \mathcal{A}$, let $M_{a,b} \in \mathcal{E}\ell(\mathcal{A})$ be the operator given by $M_{a,b}(x) = axb$. Also let L_a, R_b be the elementary operators on $\mathcal{A}$ given by $L_a(x) = ax$, $R_b(x) = xb$. Therefore, a general elementary operator on $\mathcal{A}$ is of the form $R_{\mathbf{a},\mathbf{b}} = \sum_{j=1}^{n} M_{a_j,b_j}$, where $\mathbf{a} = (a_1, \ldots, a_n)$ and $\mathbf{b} = (b_1, \ldots, b_n)$. For a Banach space $\mathcal{X}$, let $\mathcal{B}(\mathcal{X})$ (resp. $\mathcal{K}(\mathcal{X})$) denote the algebra of bounded linear (resp. compact) operators on $\mathcal{X}$. If H is a Hilbert space H, let $\mathcal{C}_2(H)$ be the ideal of all Hilbert–Schmidt operators on H.

The problems will be listed alphabetically by author.

L. Fialkow

Do there exist elementary operators other than derivations such that $\|R_{\mathbf{a},\mathbf{b}|\mathcal{S}}\| = \|R_{\mathbf{a},\mathbf{b}}\|$ for every symmetrically normed ideal $\mathcal{S}$ in $\mathcal{B}(H)$?

R. Harte

(i) Relate the conditions on an elementary operator of being hyponormal, strong Fuglede and *-hyponormal.

(ii) Let $\mathcal{A}$ be a Banach algebra and $a, b \in \mathcal{A}$. The operator $D = L_a - R_b$ satisfies the algebraic condition $D(xy) = xD(y) - D(x)y$. Similarly, one can write down an algebraic expression for $D(xyz)$.
Are there similar algebraic conditions for the operator $M_{a,b}$?

(iii) Anderson and Foias ([AF, Theorem 5.8]) show that, if $0 \neq p^* = p = p^2 \neq 1 \in A = \mathcal{B}(H)$ for Hilbert space H, then $p \in A$, and hence $L_p, R_p \in \mathcal{B}(A)$, are Hermitian while the product L_pR_p is not. Is it the case that L_pR_p is in the "Palmer subspace" Reim $\mathcal{B}(A) = \text{Re } \mathcal{B}(A) + i\text{Re } \mathcal{B}(A)$? What exactly is the numerical range of L_pR_p ?

(iv) Alternatively is there in a Banach algebra implication $ab = ba = h + ik$ with hermitian a, b, h, k implies $ab = h$ is hermitian?

[AF] J. Anderson and C. Foias, *Properties which normal operators share with normal derivations and related properties*, Pacific J. Math. 61 (1975) 133–325.

R.E. Curto and M. Mathieu (eds.), *Elementary Operators and Their Applications*, Operator Theory: Advances and Applications 212, DOI 10.1007/978-3-0348-0037-2_8, © Springer Basel AG 2011

B. Magajna

(i) Let ϕ be a mapping on $B(H)$ of the form $\phi(x) = \sum_{n=1}^{\infty} a_n^* x a_n$, where $\sum_{n=1}^{\infty} a_n^* a_n = 1$ and the sums are convergent in the strong operator topology. Suppose that H is separable and that ϕ preserves the ideal of compact operators. (For example, let a_n be orthogonal projections with sum 1.) Can ϕ be approximated pointwise in norm by a net of *completely contractive* elementary operators?

(ii) Suppose that the normal and the singular part of a complete contraction ϕ on a von Neumann algebra can both be approximated pointwise in norm by nets of completely contractive elementary operators. Does then ϕ necessarily admit such an approximation?

The answer is affirmative for completely positive maps by Corollary 4.2 in my paper in this volume. But the case of completely contractive maps can not be easily reduced to completely positive maps, perhaps not even on injective von Neumann algebras, where one could try to use the well-known 2×2 matrix technique. Namely, maps that admit pointwise approximation by elementary operators must preserve closed ideals, while the Arveson extension theorem for completely positive maps on operator systems does not automatically provide extensions with this property.

M. Mathieu

(i) Let $\mathcal{A}$ and $\mathcal{B}$ be Banach algebras or C^*-algebras. Suppose that the algebras $\mathcal{E}\ell(\mathcal{A})$ and $\mathcal{E}\ell(\mathcal{B})$ are isomorphic. Does it follow that $\mathcal{A}$ and $\mathcal{B}$ are isomorphic?

(ii) Following on from our article in this volume, characterise when a length two elementary operator $M_{a,b} + M_{c,d}$ is a spectral isometry.

P. Rosenthal

Let $\mathcal{A}$ be either $\mathcal{B}(H)$, $\mathcal{K}(H)$ or $\mathcal{C}_2(H)$, for some Hilbert space H.

(i) Characterise the equivalence classes of elementary operators arising from the relation of unitary equivalence or similarity.

(ii) Find all invariant subspaces of particular elementary operators.

(iii) Determine when a given elementary operator is subnormal or quasi-normal.

V.S. Shulman and Yu.V. Turovskii

(i) Let $a, b, a_i, b_i \in \mathcal{K}(\mathcal{X})$, $i = 1, \ldots, n$. Does the elementary operator $L_a + R_b + \sum_{i=1}^{n} M_{a_i,b_i}$ possess a "non-trivial invariant" subspace?

Here "non-trivial" means different from the closed ideals of $\mathcal{B}(\mathcal{X})$. For a more definite formulation one can consider the restriction of R to the smallest closed ideal $\mathcal{A}(\mathcal{X})$, the closure of the ideal of finite rank operators, and ask about invariant subspaces nontrivial in the usual sense, that is, different from 0 and $\mathcal{A}(\mathcal{X})$.

The problem seems to be interesting in itself as any concrete version of the invariant subspace problem. But a fresh idea leading to its solution could have applications, for example to the conjecture of W. Wojtyński [4] that a Lie algebra of Volterra operators cannot be topologically simple (see also [1]).

It should be added that for operators $R_1(x) = ax+xb$ and $R_2(x) = \sum_{i=1}^n a_i x b_i$ the existence of invariant subspaces can be easily verified: R_2 is compact, R_1 commutes with the compact operator $M_{a,b}$ (thus its non-transitivity follows from [2]). This gives a hope that the problem is not wild.

(ii) A stronger version of the previous problem is: Let $R = \sum_{i=1}^n M_{a_i,b_i}$ be an elementary operator on $\mathcal{A}(\mathcal{X})$, and for each i at least one of the operators a_i, b_i is compact. Does R have a non-trivial invariant subspace?

(iii) Let us denote, for any Banach algebra $\mathcal{A}$, by $\mathrm{Mul}(\mathcal{A})$ the norm closure in $\mathcal{B}(\mathcal{A})$ of the algebra $\mathcal{E\ell}(\mathcal{A})$ of all elementary operators.

Let $\mathcal{A}$ be a compact radical Banach algebra, for example a norm-closed algebra of Volterra (= compact quasinilpotent) operators. Is the algebra $\mathrm{Mul}(\mathcal{A})$ radical?

It should be said that the Jacobson radical of $\mathrm{Mul}(\mathcal{A})$ contains all operators $\sum_{i=1}^n L_{a_i} R_{b_i}$. Hence the semisimple quotient $\mathrm{Mul}(\mathcal{A})/\mathrm{Rad}(\mathrm{Mul}(\mathcal{A}))$ is generated by two subalgebras $\mathcal{M}_1$ and $\mathcal{M}_2$ (the images of the algebras of left and right multiplications) which consist of quasinilpotent elements and satisfy the condition $\mathcal{M}_1\mathcal{M}_2 = \mathcal{M}_2\mathcal{M}_1 = 0$. Is this possible?

Partial positive results can be found in Section 5 of [3].

[1] M. Brešar, V.S. Shulman, Yu.V. Turovskii, On tractability and ideal problem in non-associative operator algebras, (to appear in *Integral Equations and Operator Theory*).

[2] V.I. Lomonosov, Invariant subspaces for operators commuting with compact operators, *Funct. Anal. Appl.* **7** (1973), 213–214.

[3] V.S. Shulman, Yu.V. Turovskii, Tensor radical and its applications to spectral theory of multiplication operators, (*this volume*).

[4] W. Wojtyński, Engel's theorem for nilpotent Lie algebras of Hilbert-Schmidt operators, *Bull. Acad. Polon. Sci.* **24** (1976), 797–801.

E. Saksman and H.-O. Tylli

Let $\mathcal{X}$ be a Banach space. Suppose that an elementary operator $R \in \mathcal{E\ell}(\mathcal{B}(\mathcal{X}))$ is invertible. How far is the inverse R^{-1} from being an elementary operator? In particular, is it true that $R^{-1} \in \overline{\mathcal{E\ell}(\mathcal{B}(\mathcal{X}))}^{\|\cdot\|}$?

Note that, if the operator R from the last question has the form $R = L_a - R_b$, for some $a, b \in \mathcal{B}(\mathcal{X})$, then R^{-1} does belong to $\overline{\mathcal{E\ell}(\mathcal{B}(\mathcal{X}))}^{\|\cdot\|}$, see [R]. For $\mathcal{X} = \ell^2$ one may deduce from a result of Magajna [M] that R^{-1} belongs at least to the strong operator topology closed hull of $\mathcal{E\ell}(\mathcal{B}(\ell^2))$ for any invertible $R \in \mathcal{E\ell}(\mathcal{B}(\ell^2))$.

[M] B. Magajna: *A transitivity theorem for algebras of elementary operators*, Proc. Amer. Math. Soc. 118 (1993), 119–127.

[R] M. Rosenblum: *On the operator equation $BX - XA = Q$*, Duke Math. J. 23 (1956), 263–269.

Operator Theory: Advances and Applications (OT)

This series is devoted to the publication of current research in operator theory, with particular emphasis on applications to classical analysis and the theory of integral equations, as well as to numerical analysis, mathematical physics and mathematical methods in electrical engineering.

Edited by
Joseph A. Ball (Blacksburg, VA, USA), Harry Dym (Rehovot, Israel), Marinus A. Kaashoek (Amsterdam, The Netherlands), Heinz Langer (Vienna, Austria), Christiane Tretter (Bern, Switzerland)

■ **OT 216: Ruzhansky, M. / Wirth, J.**, Modern Aspects of the Theory of Partial Differential Equations (due 2011).
ISBN 978-3-0348-0068-6

■ **OT 215: Cruz-Uribe, D. / Martell, J.M. / Perez, C.**, Weights, Extrapolation and the Theory of Rubio de Francia (due 2011).
ISBN 978-3-0348-0071-6

■ **OT 214: Janas, J. / Kurasov, P. / Laptev, A. / Naboko, S. / Stolz, G. (eds.)**, Spectral Theory and Analysis. Operator Theory, Analysis and Mathematical Physics (OTAMP) 2008, Bedlewo, Poland (2011).
ISBN 978-3-7643-9993-1

■ **OT 213: Rodino, L. / Wong, M.W. / Zhu, H. (eds.)**, Pseudo-Differential Operators: Analysis, Applications and Computations (due 2011).
ISBN 978-3-0348-0048-8

■ **OT 212: Curto, R.E. / Mathieu, M.**, Elementary Operators and Their Applications (2011).
ISBN 978-3-0348-0036-5

■ **OT 211: Demuth, M. / Schulze, B.-W. / Witt, I. (eds.)**, Partial Differential Equations and Spectral Theory (2010).
Subseries Advances in Partial Differential Equations.
ISBN 978-3-0346-0508-3

■ **OT 210: Duduchava, R. / Gohberg, I. / Grudsky, S.M. / Rabinovich, V. (eds.)**, Recent Trends in Toeplitz and Pseudodifferential Operators. The Nikolai Vasilevskii Anniversary Volume (2010).
ISBN 978-3-0346-0547-2

■ **OT 209: Eisner, T.**, Stability of Operators and Operator Semigroups (2010).
ISBN 978-3-0346-0194-8

■ **OT 208: Elin, M. / Shoikhet, D.**, Linearization Models for Complex Dynamical Systems (2010).
Subseries Linear Operators and Linear Systems.
ISBN 978-3-0346-0508-3

■ **OT 207: Axler, S. / Rosenthal, P. / Sarason, D. (eds.)**, A Glimpse at Hilbert Space Operators. Paul R. Halmos in Memoriam (2010).
ISBN 978-3-0346-0346-1

■ **OT 206: Lerer, L. / Olshevsky, V. / Spitkovsky, I.M. (eds.)**, Convolution Equations and Singular Integral Operators (2010).
ISBN 978-3-7643-8955-0

■ **OT 205: Schulze, B.-W. / Wong, M.W. (eds.)**, Pseudo-Differential Operators: Complex Analysis and Partial Differential Equations (2009).
ISBN 978-3-0346-0197-9

■ **OT 204: Frazho, A. / Bhosri, W.** An Operator Perspective on Signals and Systems (2009).
Subseries Linear Operators and Linear Systems.
ISBN 978-3-0346-0291-4

■ **OT 202/203: Ball, J.A. / Bolotnikov, V. / Helton, J.W. / Rodman, L. / Spitkovsky, I.M. (eds.)**, Topics in Operator Theory. A Tribute to Israel Gohberg on the Occasion of his 80th Birthday (2010).
Volume 1: Operators, Matrices and Analytic Functions. ISBN 978-3-0346-0157-3
Volume 2: Systems and Mathematical Physics.
ISBN 978-3-0346-0160-3
Set Volumes 1/2: ISBN 978-3-0346-0163-4

MIX
Papier aus verantwortungsvollen Quellen
Paper from responsible sources
FSC® C105338

If you have any concerns about our products,
you can contact us on
ProductSafety@springernature.com

In case Publisher is established outside the EU,
the EU authorized representative is:
Springer Nature Customer Service Center GmbH
Europaplatz 3, 69115 Heidelberg, Germany

Printed by Libri Plureos GmbH
in Hamburg, Germany